Maritime Heritage in Crisis

Grounded in critical heritage studies and drawing on a Pacific Northwest Coast case study, *Maritime Heritage in Crisis* explores the causes and consequences of the contemporary destruction of Indigenous heritage sites in maritime settings. Maritime heritage landscapes are undergoing a period of unprecedented crisis: these areas are severely impacted by coastal development, continued population growth, and climate change. Indigenous heritage sites are thought to be particularly vulnerable to these changes and cultural resource management is frequently positioned as a community's first line of defense; yet, there is increasing evidence that this archaeological technique is an ineffective means of protection.

Exploring themes of colonial dislocation and displacement, Hutchings positions North American archaeology as neoliberal statecraft: a tool of government designed to promote and permit the systematic clearance of Indigenous heritage landscapes in advance of economic development. Presenting the institution of archaeology and cultural resource management as a grave threat to Indigenous maritime heritage, *Maritime Heritage in Crisis* offers an important lesson on the relationship between neoliberal heritage regimes and global ecological breakdown.

Richard M. Hutchings is a founding director of the Institute for Critical Heritage and Tourism, British Columbia, Canada. Born and raised in Seattle, Washington, he obtained his MA from Western Washington University, Bellingham, and his PhD from the University of British Columbia, Vancouver. He resides on Gabriola Island in the Salish Sea.

Archaeology and Indigenous Peoples Series
Sponsored by the World Archaeological Congress

Books in this series:

Indigenous Peoples and Archaeology in Latin America
Edited by Cristóbal Gnecco and Patricia Ayala

Indigenous Archaeologies
A Reader on Decolonization
Edited by Margaret Bruchac, Siobhan Hart, and H. Martin Wobst

Being and Becoming Indigenous Archaeologists
Edited by George Nicholas

Kennewick Man
Perspectives on the Ancient One
Edited by Heather Burke, Claire Smith, Dorothy Lippert, Joe Watkins, and Larry Zimmerman

Maritime Heritage in Crisis

Indigenous Landscapes and Global Ecological Breakdown

Richard M. Hutchings

Routledge
Taylor & Francis Group
NEW YORK AND LONDON

First published 2017
by Routledge
711 Third Avenue, New York, NY 10017

and by Routledge
2 Park Square, Milton Park, Abingdon, Oxon, OX14 4RN

Routledge is an imprint of the Taylor & Francis Group, an informa business

British Library Cataloguing-in-Publication Data
A catalogue record for this book is available from the British Library

Library of Congress Cataloging-in-Publication Data
Names: Hutchings, Richard M. (Richard Matthew) author.
Title: Maritime heritage in crisis : indigenous landscapes and global ecological breakdown / Richard M. Hutchings.
Description: Abingdon, Oxon ; New York, NY : Routledge, 2016. | Series: Archaeology and indigenous peoples series | Includes bibliographical references and index.
Identifiers: LCCN 2016025007 | ISBN 9781629583471 (hardback: alk. paper) | ISBN 9781629583488 (pbk.: alk. paper) | ISBN 9781315400020 (ebk)
Subjects: LCSH: Coastal archaeology. | Coast changes. | Cultural property–Protection. | Cultural property–Management. | Antiquities–Collection and preservation. | Indigenous peoples–Antiquities. | Indians–Antiquities.
Classification: LCC GN784 .H87 2016 | DDC 551.45/7–dc23
LC record available at https://lccn.loc.gov/2016025007

ISBN: 978-1-62958-347-1 (hbk)
ISBN: 978-1-62958-348-8 (pbk)
ISBN: 978-1-315-40002-0 (ebk)

Typeset in Bembo by
Sunrise Setting Ltd, Paignton, UK

For Marina

Epigraph

Revival, culture, heritage, environment. These are key words. ... We now find ourselves fighting a battle to turn a tide of destruction that has been rising for too many years and threatens to drown the family unit, our social structure and our environment. We have been overly self-indulgent for much too long, and it is time for change.

Change comes from understanding ourselves—our weaknesses, our strengths. That understanding can be fostered from knowledge of our past, our cultural heritage and our environment. This priceless wisdom is available from our elders, ancestors and a cultural heritage. Once we know our past, we have taken a step in understanding ourselves, and we will be able to strengthen our truths, bringing about change for the better.

Such changes can affect our many relationships—intimate ones, social and professional ones, and the one we have with our environment. These actions will help us to turn the tide, letting it wash over the land, healing those wrongs we have had a part in creating.

Roy Henry Vickers, *The Elders Are Watching*, 2009.
Reprinted with permission.

Contents

List of Figures viii
List of Tables x
List of Boxes xi
Preface xii
List of Abbreviations xv

1 The Maritime Heritage Crisis 1

2 Coastal Change 19

3 Cultural Resource Management 38

4 The *shíshálh* Coast Study 54

5 Problematizing the Heritage Crisis 88

6 Looking Forward, Looking Back 105

Appendix: The Club of Rome's Forty-Nine Critical Continuous Problems 121
References 123
Index 143

Figures

1.1	Salish Sea basin and *shíshálh* Coast study area	2
2.1	Impacts of sea level rise on steep and low-lying coasts	25
2.2	Sea levels at 2100, 2200, and 2300	26
2.3	British Columbia population, 1866–2036	34
3.1	Defining British Columbia archaeology	53
4.1	*shíshálh* Coast study area	55
4.2	Sunshine Coast population, 1956–2036	60
4.3	Three *shíshálh* Coast Study Units	64
4.4	Sechelt Study Unit non-local housing ownership on the Sunshine Coast	66
4.5	Sechelt seawall and promenade	67
4.6	Heritage protection at the waterfront	67
4.7	Angus Creek contrasts	68
4.8	Angus Creek wave erosion	68
4.9	Angus Creek heritage signage	69
4.10	Snake Bay development	70
4.11	Halfmoon Bay Study Unit	71
4.12	Halfmoon Bay recreation map	71
4.13	Smuggler Cove yachts	72
4.14	Central Pender Harbour	73
4.15	Inner Pender Harbour	77
4.16	Waterfront residences in inner Pender Harbour	78
4.17	Disturbed and destroyed sites by study unit	79
4.18	Sites disturbed and destroyed in all study unit	79
4.19	Site disturbances by category for all study unit	80
4.20	Site disturbances in Acheson and Riley's 1977 study	80
4.21	"EMPIRE"	81
4.22	Percentage of recorded sites destroyed in southern Pender Harbour	82
4.23	Site elevation	83
4.24	Percentage of recorded sites intersecting shoreline	84
4.25	Residential ownership by origin and location, 2010	85

4.26 Non-local housing ownership on the Sunshine Coast by origin, 2010 86
6.1 Indigenous heritage landscape paradox 112
6.2 Four responses to modernity 114
6.3 Issue-attention cycle 116

Tables

2.1 Major taphonomic and formation processes on California's Channel Islands 26
2.2 Determinants of archaeological site survival through inundation 28
4.1 *shíshálh* Nation goals and priority actions concerning authority over the heritage stewardship process 63
4.2 Summary comparison of three *shíshálh* Coast Study Units 78

Boxes

1.1 Global ecological breakdown in perspective: Thinking critically, part I 7
1.2 The Indigenous heritage crisis: Key readings, 1969–2014 15
3.1 The evolution of Canadian heritage policy, part I 39
3.2 The evolution of Canadian heritage policy, part II 40
4.1 Contextualizing the *shíshálh* Nation Strategic Land Use Plan 56
4.2 *shíshálh* heritage principles, part I 58
4.3 *shíshálh* heritage principles, part II 61
4.4 The Pender Harbour–Sakinaw Cultural Emphasis Area 72
4.5 *séxwʔámin* (Garden Bay) and the Pender Harbour study area 75
6.1 Global ecological breakdown in perspective: Thinking critically, part II 110

Preface

> We would do well to remember that heritage preservation and the presentation of the past are processes that erase the past just as surely as warfare, looting, or development do.
>
> (Arnold 2014: 2446–7)

In my 15 years of experience in North American archaeology, and despite all the archaeology conferences I have attended, I can honestly say I have never witnessed a meaningful public discussion about what archaeology is, or what it does. I once witnessed an attempt to start such a conversation, but learned very quickly how power is used to quash such efforts.

The 2013 Canadian Archaeology Association (CAA) plenary session, organized by Indigenous archaeologist Eldon Yellowhorn, was a panel of First Nation presenters, all speaking to their experiences with British Columbia's *Heritage Conservation Act* (HCA), the provincial legislation that governs the management of archaeological sites pre-dating 1846. The 1846 date is important because in British Columbia most settler or "historic" sites post-date this period. As such, the HCA purposefully targets Indigenous heritage, thus Indigenous people. It is in this colonial light that Yellowhorn organized his plenary session, titled Whither the *Heritage Conservation Act*: Renewal or Funeral?

Of the speakers, I was the only non-Indigenous participant. I was also the last, largely, I think, because of the title of my talk: Hard Times Bring Hard Questions: Is Archaeology Pro-Development? Is it Classist? Colonialist? Imperialist? Racist? The message from the First Nation speakers in the room that day was loud and clear: they spoke passionately about government tactics to "erase culture" by "destroying our heritage," emphasizing that "places cannot be replaced" and what was at stake was "preserving a way of life" (Hutchings and La Salle 2015a, 2015b). They asked archaeologists to help "stop development." In response to Yellowhorn's question, the panelists unanimously called for the provincial legislation's "funeral," and my paper fit right in. I argued that archaeology is a firmly colonialist project driven by capitalism to produce narratives that remain elitist, racist, and pro growth, development, and progress (Hutchings 2013). What happened next was profound.

The product of conference scheduling, a coffee break occurred just prior to my talk. During that break, all of the invited Indigenous speakers—all non-archaeologists—left the conference, a movement that radically altered the ethnic composition of the Whistler Conference Centre (as with Canada and the United States, most archaeologists in British Columbia are white). Their reason for leaving the conference, I suspect, is that archaeology's scientific renderings of the past—most often an Indigenous past—is of little interest to Indigenous people (Yellowhorn 2002), and may even be offensive (Hutchings 2012a, 2012b; McLay et al. 2008).

Those speakers who remained for questions at the end of the plenary session were Yellowhorn, another Indigenous archaeology professor, and myself. Despite the provocative and powerful subject matter, audience response to the session, and my paper, was muted, at least publicly, with one noteworthy exception. A prominent Pacific Northwest Coast archaeology professor sternly scolded me for "picking on archaeologists," noting that the issues I raised have been discussed "for decades"—that this was "old news." Thus the session abruptly ended, leaving the impression that archaeology had been absolved of any wrongdoing.

What shocked me most at that moment was how oblivious the individual seemed to be about what they were saying, all in the defense of archaeology—and after listening to two hours of public testimony by frontline Indigenous heritage practitioners and stewards about the destructive nature of archaeology. My takeaway that day was this: while it is well known among archaeologists that archaeology is pro-development, classist, colonialist, imperialist, racist, it is *not* something to be discussed in public.

It turns out that archaeologists are adamantly opposed to truth-telling when those truths concern their institution's *non-discussables*—those subjects that are important but are "so laden with anxiety and taboo" that they are only discussed in the most private and discreet locations (Barth 2013: 198–9). Archaeologists are "fearful that open discussion of these incendiary issues in polite society will cause a meltdown," for cultural resource management (CRM) is the proverbial elephant in the room: "Everyone knows this huge pachyderm is there between the sofa and the fireplace, and we go on mopping and dusting and vacuuming around it as if it did not exist."

In this way I arrived at *bureaucratic conservatism*, the position taken by administrators or bureaucrats. According to Randall Collins and Michael Makowsky (1998: 229), the principle tenet of bureaucratic conservatism is:

> "Don't rock the boat." It declares that there are no issues other than technical questions of how to get things done and does not want to take the trouble of asking what should be done and why. It simply wants to continue as usual with existing operations.

The logic behind this conservative position—a hallmark of modern institutions like the academy, archaeology, and CRM—is maintaining the status quo.

In effect, the invited Indigenous speakers at the 2013 CAA were describing the "heritage crisis" as it is playing out in British Columbia; however, their mostly white, mostly conservative audience, reflected in the comments by the vocal critic of the session and the general tenor of the room, appeared to reject the idea outright. It is my position in this book that archaeology's (pivotal) role in Indigenous heritage destruction constitutes an institutional non-discussable; thus, attempts to broach the subject are quickly shut down (and rationalized away) by those in power, who claim that such discourse is "impolite" and "incendiary." Such engagement, when it does happen, makes archaeologists "anxious."

What the invited speakers and I were proposing that spring day in 2013 is that British Columbia archaeology is complicit in the systematic destruction of First Nation heritage landscapes. Yet, because such a narrative of ecocide and ethnocide runs so *fundamentally* counter to established institutional mythology, which sees archaeologists as "saviors" of Indigenous heritage, not destroyers, the messenger(s) had to be swiftly and publicly refuted in order to preserve the institution (Kahan et al. 2011). In describing the nature of the scientific institution, Syun-Ichi Akasofu, founding director of the International Arctic Research Center, best captures my experience that day: "A scientific establishment is highly conservative and will attempt to preserve the power of its ruling group against any rebels" (Wohlforth 2004: 160). How this applies to archaeology and the Indigenous heritage crisis is a central theme of this book.

List of Abbreviations

ACRA – American Cultural Resources Association
CAA – Canadian Archaeology Association
CBA – Council for British Archaeology
CRM – cultural resource management
DAHP – Department of Archaeology and Historic Preservation (Washington State)
EPA – Environmental Protection Agency (United States)
GBPSEI – Georgia Basin-Puget Sound Ecosystem Indicators (Salish Sea)
HCA – *Heritage Conservation Act* (British Columbia)
ICOMOS – International Council on Monuments and Sites (United Nations)
JPI – Joint Programming Institute
NPS – National Park Service (United States)
PSP – Puget Sound Partnership
RCZAS – Rapid Coastal Zone Assessment Survey
SIB – Sechelt Indian Band
SIT – strategic, instrumental, and technical
UBCIC – Union of British Columbia Indian Chiefs
UN – United Nations
UNESCO – United Nations Educational, Scientific, and Cultural Organization
USGS – United States Geological Survey
WSDOT – Washington State Department of Transportation

List of Abbreviations

1 The Maritime Heritage Crisis

> The human history of coastal regions around the world has been under assault for decades, from forces that include dam building, coastal modifications, the destruction of wetlands, marine erosion, population growth and rampant development, looting, and other processes. Global warming will exacerbate the destruction of cultural resources in coastal zones through accelerated sea level rise, intensified storm cycles, and related coastal erosion.
>
> (Erlandson 2012: 137)

Indigenous maritime heritage is under assault worldwide, and archaeology is part of the problem, not the solution. In this book, I connect (*i*) the maritime heritage crisis to the global ecological crisis; (*ii*) archaeology to the maritime heritage crisis; and (*iii*) archaeology to the neoliberal state. The maritime heritage and global ecological crises are correlated insofar as they represent the same process and principle. The process is industrial capitalism, grounded in and motivated by the all-pervasive and seemingly irrepressible Western ideology of growth, development, and progress. The principle is that Western society's current system of natural and cultural heritage stewardship—that is, "resource management," including archaeology and cultural resource management (CRM)—is fatally flawed, for it works in service of the forces driving the crisis. For professional archaeologists, my claim approximates blasphemy. For Indigenous peoples, it approximates history.

To demonstrate these connections, I examine the maritime heritage crisis as it is playing out in the Salish Sea watershed, a 110,000 kilometers^2 (42,000 miles^2) basin on North America's Pacific Northwest Coast. As shown in Figure 1.1, the basin includes the major metropolitan centers of Seattle, Washington, USA, and Vancouver, British Columbia, Canada. In particular, I focus on the crisis in *shíshálh* (Sechelt) First Nation territory, located an hour-long ferry ride north of Vancouver, which has long been exploited by outsiders for its natural resources and recreational opportunities. Today, *shíshálh* First Nation territory is known to most as the "Sunshine Coast," the product of an early real estate naming ploy aimed at drawing potential homebuyers (settlers) to the area. In connecting the *shíshálh* maritime heritage crisis to Vancouver, the regional center of economic power, and Victoria, the regional center of political power, I connect it to the global economy, thus the global ecological crisis.

In Chapter One, I explain my rationale for writing this book. I introduce the global ecological crisis and define maritime heritage, linking both in the context of Indigenous heritage and the "miner's canary." I end by discussing the book's organization.

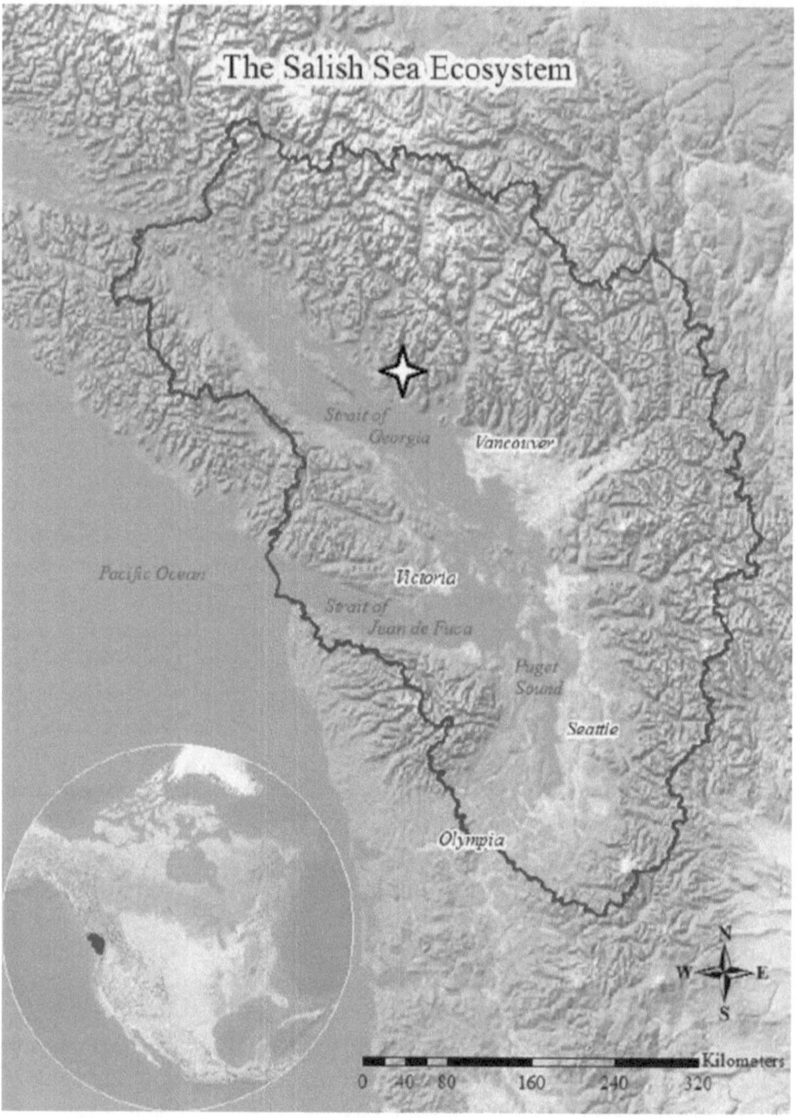

Figure 1.1 Located in North America's Pacific Northwest Coast, the transnational Salish Sea basin ecosystem includes the major economic centers Seattle, Washington (USA), and Vancouver, British Columbia (Canada), and political centers Olympia and Victoria, respectively. The *shíshálh* Coast study area, located northwest of Vancouver and denoted by the white star, is accessible only by airplane or an hour-long ferry ride.

Source: Adapted from United States Environmental Protection Agency/Environment Canada (2013).

Rationale

This study initially (and naively) set out in 2007 to "solve" the maritime heritage crisis for the Pacific Northwest Coast. The crisis, as I had originally framed it, is that sea level rise over the next century will destroy a vast number of the region's coastal archaeological sites. As a geoarchaeologist specializing in ancient coastal change, and with my experience in coastal CRM, I felt I was well-situated to quantify and explicate this "problem." Indeed, a mainstream "solution" to the crisis was rather easy to come by. Since the 1990s, European heritage specialists have developed a meticulously researched, seemingly rock-solid response.

Buttressed by the methods of scientific archaeology, humanized through maritime cultural landscape theory, and organized under the logic of CRM, I quickly concluded that the solution had been found. My research program was, for all intents and purposes, set to military precision: (*i*) phase one: Desk-top survey; (*ii*) phase two: Field survey; and (*iii*) phase three: Valuation. Then, soon after, a great unease set about me.

The disquiet emerged from the seeming normalcy and simplicity of the European model (English Heritage 2007), which has since, and not to my surprise, been imported for use in North America (Westley et al. 2011). I did not have a good grasp of why I felt it to be problematic; however, I knew it did not correlate to what I was reading and thinking at the time. Here is one example from John Bellamy Foster's (2009: 11) book *The Ecological Revolution: Making Peace with the Planet*:

> We live in a time in which the confrontation of reality with reason requires us to ask apocalyptic questions: Is the planet facing a major ecological collapse? Is civilization on the edge of the abyss? Is the survival of the human species itself in doubt? What makes the raising of these questions rational and necessary, today, is that scientific research is consistently telling us that if current trends continue, even for a century, the results for the earth and its inhabitants will be a collapse of the world as we know it.

Another example, this from Fathali Moghaddam's (2010: 1) *The New Global Insecurity: How Terrorism, Environmental Collapse, Economic Inequalities, and Resource Shortages Are Changing Our World*, vastly complicates what it means to manage resources in the late modern period:

> Hard time times bring hard questions. Calamitous wars, international terrorism, environmental degradation and global warming, interconnected global economic depression—the twenty-first century has given birth to hard times, and we now labor under hard questions. How will we manage a world severely challenged by shrinking resources, ballooning population, huge and increasing income inequalities, terrorism, torture, and environmental collapse?

Against Foster and Moghaddam, the European response seems trite because it does not allow for "hard questions" about the true scope, scale, and interconnectedness of "the problem."

Then the proverbial lightning bolt struck, setting off an irreversible chain reaction. In 2010, I acquired a copy of Raymond Rogers' (1998) *Solving History: The Challenge of Environmental Activism*, and at once my understanding of the problem began to change. "It is very clear," writes Rogers (1998: x), that heritage problems are "not about 'resolving issues'; they are about coming to terms with particular histories associated with modern economic development, which in turn, require a challenge directed at specific interests which have been the beneficiaries of that development." In light of Rogers' focus on social power, I asked: What is archaeology? What is CRM? What exactly are "resources?" And for whom, exactly, are those resources being managed? Who are the "beneficiaries?" What exactly do we mean by "management?" By "development?" What role do archaeologists and other heritage specialists/experts play in all this? Ultimately, what is "the problem?"

My fruitful engagement with *Solving History* induced me to dust off and reread my copy of Rogers' (1995) *The Oceans Are Emptying: Fish Wars and Sustainability*. There I made the second key connection, this one more fully illuminating my aforementioned unease. Rogers identifies two different ways of thinking about heritage problems (1995: 153). The first is what he calls the "strategic, instrumental, and technical approach" (what I refer to here as the "SIT" approach). This approach is typically science- and planning-oriented and "does not call the modern human project related to science, technology and capitalism into question." Rather, it focuses on "monitoring activities and reforming certain practices which are deemed to have negative consequences" for heritage (1995: 153). Known alternatively as the "problem-solving approach," the SIT (and wait) approach represents the vast majority of proposed "solutions" to heritage crises. For Moghaddam, this course of "trying to find technical solutions to moral problems is misguided" (1997: 8), yet "educators—the very people who should be able to lead us out of this path of increasing specialization—have themselves fallen victim" (2010: 7).

In opposition to the SIT approach is one based on social and cultural analysis. This alternative, the "sociohistorical approach," begins with the recognition that "the structures of everyday life and the structures that cause [heritage] problems are one and the same" (Rogers 1995: 153). It is, therefore, "all but impossible to separate what is causing [heritage] problems from the texture of a whole society." According to Rogers:

> Discussions from this perspective relate to the hegemonic domination of current realities and reject the concept that, since [heritage] problems are connected with the fabric of everyday life, it is possible to develop and maintain a regulatory perspective which can mitigate problems, since regulation itself can be part of the problem. In other words, dealing with [heritage] issues may require more than solving problems, it may be necessary

> to solve history by making everyday life problematic in order to deal with [heritage] problems.
>
> (1995: 153)

Among other things, Rogers' sociohistorical approach led me to problematize and reconsider my own "everyday life," which then, as now, is archaeology. Further, it led me to question whether archaeology and CRM might be "part of the problem."

As a result, I found it necessary to retheorize the institution of archaeology. This includes (re)defining archaeology as the sum of academic archaeology (theoretical archaeology) and CRM (applied archaeology). Because "[t]eaching is the act by which the culture of archaeology is reproduced" (Hutchings and La Salle 2014: 7), academic archaeologists are directly responsible for how the practice of cultural resource management unfolds—this, because "they alone discipline professional archaeologists in the skill-set of compliance" (Hutchings and La Salle 2015a: 20). My focus on the larger institution of archaeology/CRM opened the doors to a wide range of discourses around states, institutions, and heritage stewardship.

Given my position in these new uncharted waters, a fresh course was needed. It is at this point that I adopted a critical heritage approach (Biro 2011; Smith 2004). Rather than bridging geology and archaeology to understand the crisis, I began using critical theory, with its emphasis on social power, as a frame to better understand the history and context of the crisis. This new tack led me directly to the harbor that is modern capitalism. I have remained moored there ever since. It is a dark and stormy place.

Global Ecological Breakdown

> The nation that destroys its soil destroys itself.
>
> (Roosevelt 1937: n.p.)

Roosevelt's sentiment is a powerful one, equally applicable at both national and planetary scales. Yet, landowners the world over are currently engaged in an "orgy of soil destruction so intense that, according to the UN's Food and Agriculture Organisation, the world on average has just 60 more years of growing crops" (Monbiot 2015). Roosevelt's sentiment is not limited to just soil.

The nation that destroys its *seas* destroys itself. Jeremy Jackson and Karen Alexander (2011: 2) paint this bleak picture of the state of the marine environment:

> Since [1950], the oceans as we knew them have begun to die. Most of the largest fish are gone and, according to the latest conservative estimate, more than 80 percent of the world's major industrial fisheries have crashed or are over- or fully exploited. Sports fishers pay more and more money to catch

> fewer and smaller fish. Apex predators like tuna, salmon, and swordfish—and the people who eat them—are increasingly full of mercury, dioxins, and PCBs. Gigantic amounts of plastic are trapped in ocean gyres, and dead zones of hypoxic waters have increased from a few dozen in the 1950s to more than four hundred today. Reef corals are dying en masse from outbreaks of bleaching and disease fueled by rising temperatures, and the acidification of surface waters due to increased carbon dioxide threatens virtually all sea life with calcareous skeletons, including corals, shellfish, and plankton.

As a result, "entire ecosystems are in danger of extinction." Estuaries and coastal seas, along with coral reefs, are "critically endangered."

In this book, and in light of Box 1.1, I use the term *global ecological crisis* to refer to the socioenvironmental consequences of the acceleration in human activity from the start of the industrial revolution in 1750 to the present day (International Geosphere–Biosphere Project 2015; Steffen et al. 2015). As the most dramatic changes have occurred since 1950, that year marks the beginning of the *Great Acceleration*, the period when major planetary changes became directly linked to the global economic system:

> The Great Acceleration marks the phenomenal growth of the global socio-economic system, the human part of the Earth System. It is difficult to overestimate the scale and speed of change. In little over two generations—or a single lifetime—humanity (or until very recently a small fraction of it) has become a planetary-scale geological force. Hitherto human activities were insignificant compared with the biophysical Earth System, and the two could operate independently. However, it is now impossible to view one as separate from the other.
>
> (Steffen et al. 2015: 93–4)

The Great Acceleration corresponds thematically and temporally with *late modernity*. An extension of modernity, it is defined by "complex, global capitalist economies and a shift from state support and welfare to the private provision of services" (Harris 2004: 3). According to James C. Scott (1998: 4), late modernity is best thought of as:

> a strong, one might even say muscle-bound, version of the self-confidence about scientific and technical progress, the expansion of production, the growing satisfaction of human needs, the mastery of nature (including human nature), and, above all, the rational design of social order commensurate with the scientific understanding of natural laws.

A by-product of unprecedented scientific and industrial expansion in the West, the late modern ideology is "unfaltering confidence in science and technology as means to reorder the social and natural world" (Scott 1998: 4).

The global ecological crisis, defined here in terms of the Great Acceleration and late modernity, is the overarching socioenvironmental context for the maritime heritage crisis. As such, I ask how the two crises are related, and what that might mean for the future. As a baseline for that exploration, I identify here three key aspects of the global ecological crisis. The first is the core tenet of the Great Acceleration: that human activity, predominantly global capitalism, is now the prime driver of planetary change (Bodley 2008a; Foster et al. 2010). The second is that by destroying essential human life-support systems, both natural and cultural, the global ecological crisis negatively impacts human health and wellbeing (Dolan and Ommer 2008; Millennium Ecosystem Assessment 2005a, 2005b).

Third, while the global crisis affects everyone, it does not affect everyone *equally* (Bodley 2008b). As I demonstrate in this book, Indigenous peoples are particularly vulnerable to coastal change, and their heritage landscapes have been, and continue to be, severely impacted. Because of the relationship between Indigenous peoples, their heritage places, and the survival of their culture, they face an especially difficult future in light of the maritime heritage crisis.

Box 1.1 Global ecological breakdown in perspective: Thinking critically, part I

Only within the moment of time represented by the present century has one species—man—acquired significant power to alter the nature of the world.

Rachel Carson, 1962: 5

The world is experiencing a global ecological crisis, one that is both a product of past ecological and economic patterns and a challenge for the future. … [P]erceptions of planetary destruction and calls for the Earth's renewal abound. Can planetary life sustain itself in the face of industrial assaults?

Carolyn Merchant, 1992: 17–18

A profound transformation of Earth's environment is now apparent, owing not to the great forces of nature or to extraterrestrial sources but to the numbers and activities of people—the phenomenon of global change. Begun centuries ago, this transformation has undergone a profound acceleration during the second half of the 20th century. During the last 100 years human population soared from little more than one to six billion and economic activity increased nearly 10-fold between 1950 and 2000.

Will Steffen et al., 2004: 2

Today humanity uses the equivalent of 1.5 planets to provide the resources we use and absorb our waste. This means it now takes the Earth one year and six months to regenerate what we use in a year. Moderate

UN scenarios suggest that if current population and consumption trends continue, by the 2030s, we will need the equivalent of two Earths to support us. And of course, we only have one. … The result is collapsing fisheries, diminishing forest cover, depletion of fresh water systems, and the buildup of carbon dioxide emissions, which creates problems like global climate change.

Global Footprint Network, 2015: n.p.

The more than 7 billion humans alive right now are, by virtue of our very existence, contributing to a problem of a planetary scale—and we could number as many as 5 billion more by this century's end. We're taking up so much space, and so quickly, that the planet lost half of its total population of wildlife in just four decades. Climate change, spurred on by our voracious appetite for fossil fuels, is threatening Earth's very future.

Lindsay Abrams, 2015a: n.p.

The sixth mass extinction is nearly upon us. Species on Earth are dying out at a rate one thousand times greater than they were before humans began altering the environment. By the end of this century, scientists warn, anywhere from 20 to 50 percent of the species on Earth could be lost forever.

Lindsay Abrams, 2015b: n.p.

The globally averaged temperature over land and ocean surfaces for 2014 was the highest among all years since record keeping began in 1880. The December combined global land and ocean average surface temperature was also the highest on record.

National Oceanic and Atmospheric Administration, 2015: n.p.

We are in the midst of a global ecological crisis. And yet, like Nero, we fiddle while Rome burns. Global warming is happening. Human population is growing. Land and water supplies are used and depleted at an ever-expanding rate. Species and habitats are destroyed and biodiversity is lost. Pollution and toxic waste pile up. Despite several decades of acute awareness of these ecological problems, we have made little progress toward sustainable solutions.

Andrew Fiala, 2010: 51–2

The alarm bells of the climate crisis have been ringing in our ears for years and are getting louder all the time—yet humanity has failed to change course. What is wrong with us?

Naomi Klein, 2015: n.p.

The seriousness of the ecological crisis lays bare the depth of man's moral crisis.

Pope John Paul II, 1990: n.p.

Maritime Heritage

> The Sea is History.
>
> (Walcott 1979: n.p.)

Maritime heritage is heritage that is *of* or *pertains to* the sea (Casteel and Quimby 1975: 1). The succinctness of this definition belies the fact that no agreed upon definitions exist for "the sea." Discussing its representation in literature, Jonathan Raban writes that the sea is "not a verifiable object, to be described, with varying degrees of success and shades of emphasis, by writers of different periods; it is, rather, the supremely liquid and volatile element, shaping itself newly for every writer and every generation" (1992: 3). A similar conclusion is reached by John Mack in his cultural history of the sea: in the end, there exist as many "different seas" as there are different "concepts of the sea" (2011: 36, 72). There is no accepted technical definition of "sea" among oceanographers either (the term is interchangeable with "ocean"). As such, we might minimally agree that the sea is a large body of salt water that (*i*) is part of surrounding waters or (*ii*) is surrounded in part or in whole by land.

Seas are an important and sometime overlooked facet of the human experience (Cunliffe 2008; Erlandson 2001; Erlandson et al. 2015). Not only do they provide a significant portion of the plants and protein in the human diet, they greatly facilitate long-distance trade, travel, and migration, especially when reliable watercraft are present. Shores and the terrestrial land adjacent to seas—collectively referred to here as "coasts"—are central places for human activity and settlement, and thus for human culture (Turner et al. 2003). As such, humans have long played an important role in shaping the coastal landscape, for better or for worse (Erlandson and Rick 2008; Schwartz 2005).

Heritage

Like the sea, there is little consensus around what exactly constitutes "heritage." Minimally, it refers to that which is "inherited." In this regard, *heritage* is closely linked to—if not synonymous with—"culture" and "nature," and thus "environment" (Barnett and Casper 2001; Barry 2007) and "history" (Barman 2005; Cronin 2011; Nash 2014; Spence 1999). Culture is, according to E. B. Tylor, the founder of cultural anthropology, "that complex whole which includes knowledge, belief, art, morals, law, custom, and any other capabilities and habits acquired by man as a member of society" (1974 [1871]: 1). Tylor's definition of culture applies equally well to cultural heritage.

While precise or narrow definitions of heritage are inherently flawed, they are nevertheless regularly used today, especially for the purposes of heritage or resource management. Examples include distinguishing between "natural" (non-human) and "cultural" (human) heritage, "tangible" (archaeological) and "intangible" (non-archaeological) cultural heritage, and "prehistoric" (Indigenous) and "historic" (typically settler) cultural heritage. Such divisions fail even the most basic of tests. In the context of the Pacific Northwest Coast (Deloria 2012), for example, do salmon constitute "natural" or "cultural" heritage? Are salmon a "natural" or "cultural" resource? The answer is they are always *both*.

Despite how illogical such divisions seem to be, resource managers nevertheless carry out daily such distinctions in regards to Indigenous culture/heritage/people, as a matter of practice. Heritage is typically divided into three major categories or "silos": (*i*) natural heritage, (*ii*) cultural heritage, and (*iii*) intangible cultural heritage (Dearborn and Stallmeyer 2010: 15–23). As delineated by the United Nations (UN) in their 1972 Convention Concerning the Protection of the World Cultural and Natural Heritage (United Nations Educational, Scientific, and Cultural Organization [UNESCO] 1972: 2), *natural heritage* includes physical, geological, and biological formations or groups, including plant and animal habitats. In contrast, *cultural heritage* includes monuments, architecture, and other human-constructed sites.

The idea of "intangible" cultural heritage is a more recent conceptualization. The UN, in their 2003 Convention for the Safeguarding of the Intangible Cultural Heritage (UNESCO 2003: 2), defines *intangible cultural heritage* as "living heritage," a category that includes:

> the practices, representations, expressions, knowledge, skills—as well as the instruments, objects, artefacts and cultural spaces associated therewith—that communities, groups and, in some cases, individuals recognize as part of their cultural heritage. This intangible cultural heritage, transmitted from generation to generation, is constantly recreated by communities and groups in response to their environment, their interaction with nature and their history, and provides them with a sense of identity and continuity, thus promoting respect for cultural diversity and human creativity.

Since the 1970s, observers have increasingly recognized that cultural and natural heritage are inherently interconnected, and their preservation "cannot be considered independently" (Dearborn and Stallmeyer 2010: 15). Likewise, there is now recognition of a "deep-seated interdependence between the intangible cultural heritage and the tangible cultural and natural heritage" (UNESCO 2003: 1; see also Andrews and Buggey 2012; Buggey 1999, 2004; Prosper 2007).

Indigenous Heritage Landscapes

Susan Buggey (2004) and Lisa Prosper (2007) use the term "Aboriginal cultural landscape" to capture the holistic nature of Indigenous heritage. I use the term "Indigenous heritage landscape" in this book to refer to the same concept. While the concept of "Indigeneity" is complex (Clifford 2007; Ross et al. 2011: 21–4), I use the term here in its most straightforward formulation: "'Indigenous' peoples are 'native' to a particular place, original to their lands rather than having emigrated from elsewhere" (Ross et al. 2011: 21). Many Indigenous histories tell how they "have either been created by or from the land, or to have themselves created their territories."

For Buggey (1999: 43), Aboriginal cultural landscapes are a way of approaching Indigenous history that:

> focuses upon the complex relationship that Aboriginal people have with the land. They are not relicts but living landscapes—the cosmological, mythological, and spiritual world of the people associated with them as well as the environment of the day-to-day activities of living on the land.

Indigenous traditional knowledge, passed through oral tradition from generation to generation, "connects these spiritual relationships to the land through narratives, place names, sacred sites, rituals, and behavior patterns that are tied to the spirits of the land" (Buggey 1999: 43).

An *Indigenous heritage landscape* is, therefore, a "place valued by an Aboriginal group (or groups) because of their long and complex relationship with that land. It expresses their unity with the natural and spiritual environment. It embodies their traditional knowledge of spirits, places, land uses, and ecology" (Buggey 1999: 27). Significantly, for archaeologists, material remains of these associations "may be prominent, but will often be minimal or absent."

Indigenous Maritime Heritage Landscapes

In developing the concept of an Indigenous maritime heritage landscape, I fuse the aforementioned ideas about Indigenous heritage landscapes with archaeologist Christer Westerdahl's (1992) notion of a "maritime cultural landscape." Westerdahl coined the term to capture the holistic nature of "maritime culture on land as well as underwater" (1992: 5). The concept aids in placing local landscapes in broader contexts, such as a much larger interaction sphere, and signifies the "human utilization (economy) of maritime space by boat: settlement, fishing, hunting, shipping and its attendant subcultures, such as pilotage, lighthouse and seamark maintenance." It comprises the entire network of marine routes, "old as well as new, with ports and harbours along the coast, and its related constructions and remains of human activity, underwater as well as terrestrial. In this way it mirrors the entire range of maritime economies, that is, mariculture" (1992: 6). Again, while archaeology emphasizes material culture, a vital part of the Indigenous maritime cultural landscape is "immaterial, cognitive, or indicatory" (Westerdahl 1992: 6).

An *Indigenous maritime cultural landscape* is, therefore, a coastal place valued by Indigenous people as a result of their enduring and complex relationship with that land/sea. Following Buggey, it expresses a maritime culture's unity with the environment, embodying its traditional knowledge and ecology (Berkes 2008; Ross et al. 2011). Following Westerdahl, while Indigenous maritime cultural landscapes are defined by both tangible and intangible elements, place names are of greatest import. In this regard, and not including lived experiences (Ingold 2011), place-name studies (e.g. Rozen 1985) and studies that rely heavily on place names and traditional knowledge (e.g. Carlson 2001; Cruikshank 2005) most closely reflect the Indigenous maritime cultural landscape. By all appearances, archaeology has limited utility when it comes to translating Indigenous maritime cultural landscapes.

The Maritime Heritage Crisis

> The next decades hold a great deal of uncertainty for many cultural resources throughout the world, particularly those in the coastal zone.
>
> (Caffrey and Beavers 2008: 94–5)

> Heritage sites are a vital part of our First Nations communities.
>
> (Union of British Columbia Indian Chiefs [UBCIC] 2013: 12)

Globally, heritage in the coastal zone is under enormous pressure from industrial-scale resource extraction and development, as well as coastal tourism. Infinitely complicating matters is that rising seas over the coming centuries will drown those landscapes not already ravaged by development. Of concern here is that for Indigenous communities, heritage landscapes constitute a "vital part" of culture (Buggey 1999; UBCIC 2013), and are recognized by the UN (2008) as "essential to cultural survival." For this reason, the global scale destruction of maritime heritage constitutes a major threat to living coastal cultures.

Coastal change is a natural process insofar as coasts and coastlines have always been changing. Sea levels have risen and fallen for as long as there have been seas, cyclically drowning and exposing land masses according to the laws of celestial physics. Bluffs, beaches, and coastlines have always been in flux, following those same laws. However, coastal change has increasingly become a cultural process, wherein humans are the driving force behind change.

An important by-product of the human dimension is that the term "coastal change" in many ways defies definition. Perceptions of "coastal change" vary between individuals, places, and times. What matters, however, is why the issue is important, and Rosemary Ommer, along with the Coasts Under Stress Research Project Team, provide a concise explanation in their book *Coasts Under Stress* (2007: 3):

> Changes in natural environments have interacted with political, industrial, and social change to adversely affect the health of the people who live there, their communities, and the natural environments in which they are embedded. … We do not have enough resources left, even if we stop degrading ecosystems now, because already almost two-thirds of our resources are used up. We need to understand what has happened, what is going on and why … and then focus on recovery. In short, Canada and the wider world need to make urgent choices. Our accustomed patterns drove us into this mess, and we need new ideas, new ways of thinking, to get us out.

The issue of coastal change thus extends far beyond physical landscape change to include a complex web of factors and processes. It also highlights the human toll of such changes.

For the purposes of constraining this inquiry, the term "coastal change" is used narrowly to refer to just two spheres: sea level rise and population growth. These processes combined arguably pose the greatest threats to coastal heritage, today and in the foreseeable future (Erlandson 2012; Gurran et al. 2007). Discussed in detail in Chapter Two, here I provide a brief overview of coastal change, which defines the maritime heritage crisis.

Sea Level Rise

Anthropogenic or human-induced sea level rise, an outcome of climate change, "may result in more frequent flooding and salt contamination of low lying areas, as well as storm damage to areas and structures previously above high water" (BC 2007: 95). Craig Landry (2011: 259) provides this overview for the United States' eastern seaboard:

> The coastal zone is one of the most dynamic natural systems on earth, with unremitting wind and waves, occasional storms, and sea level change playing key roles in process and evolution. ... Climate change threatens to increase the intensity of coastal storms and accelerate sea level rise. Analysis conducted by the Heinz Center (2000) suggests that one in four homes within 500 feet of the U.S. east coast could be directly or indirectly lost to erosion in the next 60 years, at a potential cost of $530 million each year.

Not surprisingly, impacts on coastal heritage are expected to be severe. As Jon Erlandson (2012: 139) puts it, "rising seas, accelerated coastal erosion, and larger and more frequent mega-storms will destroy many of the world's most important coastal archaeological sites." While heritage destruction is already a major problem, "sea level rise of one or two meters in the decades to come will increase these problems by an order of magnitude." Compounding the problem is that while coastlines are dynamic, archaeological sites are "fixed" (Murphy et al. 2009: 9).

Considering the human dimension, it is projected that adverse environmental changes will "likely challenge the health and wellbeing of coastal communities already under considerable stress. Researchers anticipate multiple direct and indirect community and population health consequences from climate change" (Dolan and Ommer 2008: 27). An example of this is the rapidly growing environmental refugee crisis (Bardsley and Hugo 2010; Bates 2002). As Myers pointed out in 1997, "[e]nvironmental refugees could become one of the foremost human crises of our times" (1997: 181). This issue is of great concern in coastal zones as they "may feature as many as five billion people, or two-thirds of all people in the world, making them specially susceptible to sea level rise and associated problems" (1997: 173). As a result, climate change has been linked with increased violence, and governments around the world are actively preparing, including the United States and Canada.

Population Growth and Change

Worldwide, the "twin forces" of rising affluence and population are rapidly transforming coastal communities (Gurran et al. 2007: 445). Coastal urbanization, especially in undeveloped, nonmetropolitan contexts, constitutes "one of the most significant challenges" for coastal management (2007: 446). The environmental and socioeconomic repercussions of amenity-driven migration into coastal areas are "significant," and the social impacts are "profound" (2007: 451). Unlike other high-growth contexts, usually in major urban centers, "population growth in undeveloped coastal areas occurs in extremely fragile environmental and cultural landscapes." Many local communities are "unprepared for the sudden increases in demand for new housing and tourist developments in their localities and have inadequate controls in place" (2007: 451).

Discussed further in Chapter Two, the impact of population growth and coastal sprawl on heritage landscapes is significant, and the primary context for archaeology/CRM.

Resource Management

While the concomitant issues of rising human population, growing consumption, and increasing resource exploitation—collectively, the major historic drivers of coastal change—have been intensively studied for over 25 years (Schwartz 2005), there is little agreement on how to actually enact positive change. For many reasons, there has been and continues to be disagreement at every turn. There is disagreement as to what actually constitutes the "coast" (Graber 2005; Oertel 2005), assertions that more and better science is the solution (McFadden 2007), counter-assertions that that science is not the "limiting factor" (Billé 2008), and many calls for more "integrated" approaches (Lertzman 2009; Ng'ang'a 2006). While some view the issue through an economic lens (Landry 2011), others take a legal perspective (Cutting et al. 2011) or call for more humanistic approaches (Thorburne 2000). While some look to "bottom-up" or local/community-based solutions (McGinnis and McGinnis 2011; Thorburne 2000), others promote "top-down" management models (Ng'ang'a 2006; Pilkey and Young 2009). Some heritage experts call for a shift in focus from "resources" to "heritage" (Smith 2006), and some have considered removing archaeology from heritage altogether (Waterton and Smith 2009).

Today, willingly or not, coastal communities look to government leaders, public policy-makers, coastal management experts, coastal resource managers, and archaeologists to solve late modern coastal change problems. As a consequence, mainstream discourse surrounding the crisis is largely centered on and anchored in the language, ideology, and bureaucracy of "resource management"; however, such resource management has itself been implicated as part of the problem.

In his analysis of "our unprotected heritage," Thomas King (2009: 7) considers the management process to be "pretty much a sham." The general consensus among those who study the resource management problem is that (*i*) the system is fundamentally broken and (*ii*) by most accounts things are getting much worse, not better. These issues are discussed in more depth in Chapter Three when I review the history of resource management in British Columbia. They are also central to my *shíshálh* Coast study, discussed in Chapters Four and Five.

The Miner's Canary

> Indigenous peoples are like the miner's canary—when their cultures and languages disappear, this reflects the profound sickness in the ecology. Biological and cultural diversity are basic indicators of ecological health.
>
> (Havemann and Whall 2002: iii)

The maritime heritage crisis is not unique. Rather, it represents the coastal expression of the larger global heritage crisis, whereby unsustainable heritage management means that communities must come to accept significant loss of heritage (Barthel-Bouchier 2013: 103–27). Similarly, the Indigenous heritage crisis is not without corollary and precedence (Allen 2010). As illustrated in Box 1.2, there is a long history of discourse surrounding the Indigenous heritage crisis. In my experience, however, these foundational texts are not part of archaeological discourse because they do not constitute "archaeology" (i.e. they are not about the physical remains of the past). This division defines archaeology's silo.

Box 1.2 The Indigenous heritage crisis: Key readings, 1969–2014

- Vine Deloria, Jr. (1969) *Custer Died for Your Sins: An Indian Manifesto*. New York: MacMillan. [See especially Chapter Four, "Anthropologists and Other Friends."]
- John H. Bodley (2008a [1975]) *Victims of Progress*. 5th Edition. Lanham, MD: AltaMira Press.
- Jack D. Forbes (2008 [1978]) *Columbus and Other Cannibals: The Wetiko Disease of Exploitation, Imperialism, and Terrorism*. Revised Edition. New York: Seven Stories Press.
- M. Annette Jaimes (ed.) (1992) *The State of Native America: Genocide, Colonization, and Resistance*. Boston, MA: South End Press.
- Roger Echo-Hawk and Walter R. Echo-Hawk (1994) *Battlefields and Burial Grounds: The Indian Struggle to Protect Ancestral Graves in the United States*. Minneapolis, MN: Lerner Publishing Group.

- Donald A. Grinde and Bruce E. Johansen (1995) *Ecocide of Native America: Environmental Destruction of Indian Lands and Peoples.* Santa Fe, NM: Clear Light Publishers.
- Vine Deloria, Jr. (1997) *Red Earth, White Lies: Native Americans and the Myth of Scientific Fact.* New York: Scribner.
- Mark David Spence (1999) *Dispossessing the Wilderness: Indian Removal and the Making of the National Parks.* New York: Oxford University Press.
- Marie Battiste and James (Sa'ke'j) Youngblood Henderson (2000) *Protecting Indigenous Knowledge and Heritage: A Global Challenge.* Saskatoon: Purich Publishing.
- Mario Blaser, Harvey A. Feit, and Glenn McRae (2004) *In the Way of Development: Indigenous Peoples, Life Projects and Globalization.* London: Zed Books.
- Ian J. McNiven and Lynette Russell (2005) *Appropriated Pasts: Indigenous Peoples and the Colonial Culture of Archaeology.* Lanham, MD: AltaMira Press.
- Thomas D. Hall and James V. Fenelon (2009) *Indigenous Peoples and Globalization: Resistance and Revitalization.* Boulder, CO: Paradigm.
- Taiaiake Alfred (2009) *Peace, Power, Righteousness: An Indigenous Manifesto.* 2nd Edition. New York: Oxford University Press. [See especially Part Two, "Power."]
- Lynda V. Mapes (2009) *Breaking Ground: The Lower Elwha Klallam Tribe and the Unearthing of Tse-whit-zen Village.* Seattle, WA: University of Washington Press.
- Zoltán Grossman and Alan Parker (eds.) (2012) *Asserting Native Resilience: Pacific Rim Indigenous Nations Face the Climate Crisis.* Corvallis, OR: Oregon State University Press.
- Sharon J. Ridgeway and Peter J. Jacques (2014) *The Power of the Talking Stick: Indigenous Politics and the World Ecological Crisis.* Boulder, CO: Paradigm.

My premise in *Maritime Heritage in Crisis* is that Indigenous heritage landscapes are a proxy for global ecological health, a fundamental concern of all people. One reason this relationship exists, especially in colonial contexts like Canada and the United States, is that Indigenous lands and people are colonized lands and people; that is, in an imperial formula, one is superimposed on top of the other. Consider, for example, that to pollute a Native American stream is to pollute an American stream. To deforest a First Nation coast is to deforest a Canadian coast. In this way, what happens to Indigenous people and their heritage reflects back on the colonial culture, physically and psychologically. This is the premise of the miner's canary metaphor (Cohen 1953).

Writing about Indigenous peoples and economic development, Paul Havemann and Helena Whall (2002: 10–11) use the miner's canary metaphor to link the local destruction of cultural heritage to global ecological breakdown. They note that "coerced dispossession of Indigenous peoples from their lands and sacred sites and their forced assimilation into dominant societies has accelerated since Columbus reached the western hemisphere in the 15th century, causing ethnocides and genocides throughout the world." As a consequence:

> Indigenous peoples are amongst the most vulnerable peoples in the world. Today ethnocide through the loss of the ecological contexts that keep Indigenous cultures alive can be measured in massive assaults on human well being, loss of traditional knowledge, loss of language and loss of self determination. Contemporary globalization, premised by a market first ideology, exacerbates ethnocide of Indigenous peoples as well as posing a new threat of ecocide, never before experienced by human kind. Unsustainable development is already impacting negatively upon all the Earth's people, other species and the ecosystems. Indigenous peoples are already among the first people to sound the early warning that this threat is turning into a brutal reality.
>
> (Havemann and Whall 2002: 10–11)

That the vast majority of North American archaeology is performed in the context of development as CRM makes the miner's canary metaphor all the more relevant.

Book Organization

In this book, I use the miner's canary metaphor to illustrate how the destruction of local Indigenous maritime heritage landscapes points to underlying cultural problems that affect everyone, not just Indigenous peoples. I suggest that how archaeologists respond to the maritime heritage crisis can tell us a lot about the nature of global ecological breakdown.

In Chapter Two, I focus on the subject of coastal change, emphasizing the physical impacts of development and climate change on Indigenous maritime heritage landscapes. This includes close examination of coastal sprawl and sea level rise, with an emphasis on the transboundary Salish Sea basin.

In Chapter Three, I discuss responses to coastal change, characterizing resource management as the mainstream reaction. This is put into historical context with an emphasis on British Columbia resource management, including archaeology. Analysis of the literature demonstrates the problematic relationship between archaeology/CRM and development.

In Chapter Four I present my study of coastal change in *shíshálh* First Nation territory. I provide background information on the *shíshálh* Nation and their approach to heritage stewardship, and present the results of my field survey.

Past, present, and future impacts are considered, and the extent of heritage landscape destruction is quantified.

In Chapter Five, I problematize the maritime heritage crisis, using the power-elite dynamic to explicate Indigenous heritage destruction. I discuss archaeology/CRM as ideology and tool of the state, designed to clear Indigenous peoples from the landscape in advance of capitalist development. Resulting in slow violence and solastalgia, archaeology is thus part of the problem, not the solution to the maritime heritage crisis.

In Chapter Six, I conclude by "looking forward" and "looking back." I provide an overview of the book and revisit the miner's canary.

2 Coastal Change

My goal in Chapter Two is to establish a detailed baseline from which to evaluate coastal change and the maritime heritage crisis. Toward that end, I attend here to the twin problems of sea level rise and coastal sprawl. This sets the stage for both Chapter Three, where I discuss resource management as the mainstream response, and Chapter Four, where I apply the coastal change concept to the maritime heritage crisis as it is unfolding in coastal British Columbia.

Sea Level Rise

Late modern sea level rise is one of the "most robust manifestations" of anthropogenic climate change (Watson et al. 2015), the result of increased deforestation and fossil-fuel-related greenhouse gas emissions since the industrial revolution (1750–present). By anthropogenic climate change I mean a change of climate that is attributable, directly or indirectly, to human activity that alters the composition of the global atmosphere. In this regard, the sea level rise threat is not natural but cultural.

General Threats

Because sea level rise and other climate-related coastal changes are so often characterized as "threats," they can all too easily be misconstrued as future problems. Rather, climate change is actively impacting North American coasts. Harford et al. (2010: 4) are clear that "our climate is changing and the impacts are already being felt across Canada." They identify seven categories of climate change threats, all with relevance to coastal communities, thus maritime heritage. Those categories are:

- increased severity and frequency of heat events;
- severe storms, including extreme wind, rain, ice, and/or snowfall;
- wildfires;
- water shortages and drought;
- changes in the cryosphere—melting permafrost, sea ice, lake ice, and snow;

- shifting ranges and altered ecosystems, including disease, pest and invasive species migration; and
- sea level rise, storm surges, coastal and shoreline erosion.

Climate change impacts on natural heritage have been considered for far longer and in greater depth than for cultural heritage. Focusing on sea level rise and biodiversity on United States coastlines, Walter Reid and Mark Trexler (1991: 2) describe the drowning of natural heritage this way:

> As sea levels rise, coastal erosion and the severity of coastal flooding will increase, and coastlines will recede unless stabilized by dikes or through sand nourishment. Salt-water intrusion into groundwater, rivers, bays, and estuaries will increase. Changes in rainfall patterns and temperature will modify salinity gradients in estuaries and alter rates of river delta sedimentation, and coastal currents and upwelling patterns are likely to shift geographically and change in intensity. All of these "sea changes" will affect biodiversity in the nation's coastal zones.

Sea level rise will substantially reduce coastal wetland area and exacerbate threats to coastal species. "Virtually nothing can be done," say Reid and Trexler (1991: 38), "to mitigate some of the impacts of climate change on some coastal ecosystems."

There is a rapidly growing body of literature around cultural heritage and climate change (Australian National University 2009; Barthel-Bouchier 2013; Berenfeld 2015; Eyring and Goeken 2015). Impacts of climate change on archaeological heritage, as identified in the World Heritage Centre Expert Report (Cassar et al. 2006) and summarized by Susan Shearing (2007: 9–10), include:

- impacts upon the level of survival of archaeological evidence due to changes to the hydrological, chemical, and biological processes of soil;
- damage resulting from increases in soil moisture and salt which are absorbed through the porous structure of historic buildings leading to crystallisation on surfaces through drying;
- increased infestation of timber and other organic building materials by pests migrating from other altitudes and latitudes;
- damage resulting from flooding, increased intensity and frequency of storms and wind gusts; and
- desertification, salt weathering, and erosion threats to cultural heritage in desert areas; and risks to movable cultural heritage through higher levels of humidity, temperatures, and ultraviolet light.

The impacts of climate change on cultural heritage are far more complicated than for archaeological heritage (e.g. Barthel-Bouchier 2013) because they involve living human beings and their emotions, as discussed later in this book in relation to sea level rise. In 2009, the European Joint Programming Institute (JPI) examined European and global challenges related to climate change, cultural heritage, and security (JPI 2009: 1). One key finding is that "[m]ost European citizens live in or around cultural heritage with which they identify closely because of the uniqueness and irreplaceable value of historic buildings, collections and sites for their tangible and intangible values." This demonstrates the "central role that cultural heritage has in the lives of communities," yet climate change "threaten[s] the existence of the cultural environment of Europe." Before these combined risks become irreversible threats, the authors of the report conclude, "concerted action is needed to protect, strengthen and transform the unique historic environment of Europe."

Reviews of climate change threats to heritage in coastal Northern America can be found in Caffrey and Beavers (2008), Blankholm (2009), and Goetz (2010: 6–21). In the following sections, I highlight how these issues are relevant to the Pacific Northwest Coast.

Washington State

Climate change will trigger significant physical and chemical stressors in coastal Washington State (Huppert et al. 2009: 285), including: (*i*) inundation of low-lying areas by high tides as sea level rises; (*ii*) flooding of coasts during major storm events, especially near river mouths; (*iii*) accelerated erosion of coastal bluffs; (*iv*) shifting of beach profiles, moving the position of the Mean High Water line landward; (*v*) saltwater intrusion into coastal freshwater aquifers; and (*vi*) increased ocean temperature and acidity. While similar forces will be operating everywhere, shore areas will respond differently depending upon substrate, slope, and the surrounding conditions. As beaches respond to sea level rise, coastal property lines and intertidal aquaculture leases will need to be (re)defined through modified property laws.

As sea level rise pushes shorelines and tides closer to homes and infrastructure, coastal development will be threatened (Huppert et al. 2009: 296). Major Washington ports and harbors will be seriously impacted, including the Ports of Seattle, Tacoma, Everett, Olympia, Grays Harbor, and Port Angeles, as well as the many smaller ports and marinas. Rising seas will also impact Washington State's commercial shellfish industry, estimated to be approximately US$75 million a year (2009: 299).

Puget Sound's shoreline, estimated at 2,411 kilometers (1,477 miles) in length has many facilities and residences that will be impacted by sea level rise (Huppert et al. 2009: 290). Approximately 90 percent of Puget Sound's shorelines have single-family residences or are available for residential development (Taylor et al. 2005: 1). Rising seas—and responses to it, notably increased armoring (Shipman et al. 2010a, 2010b)—will have a tremendous impact on already heavily degraded Puget Sound coastal ecosystems (Dethier 2010; Gelfenbaum et al. 2006; Georgia Basin-Puget Sound Ecosystem Indicators [GBPSEI] 2002, 2006; Glick et al. 2007; Puget Sound Partnership [PSP] 2011).

British Columbia

In 2011, the Flood Safety Section of the British Columbia Ministry of Forests, Lands and Natural Resource Operations published technical guidelines for coastal land management in light of future sea level rise (Ausenco Sandwell 2011a; see also Ausenco Sandwell 2011b, 2011c; Kerr Wood Leidal 2011). Those guidelines make provision for a sea level rise of 0.5 meters by the year 2050, 1.0 meter by 2100, and 2.0 meters by 2200, with an expectation of seas reaching 3.0 meters by 2300 (BC 2013a, 2013b). Given that governments often understate climate threats, and because sea level rise is rapidly accelerating (Watson et al. 2015), it is only because the estimates correlate with current expert assessments for the period 2100–2300 (Horton et al. 2014) that they are used here as the baseline for the *shíshálh* Coast study. I will return to the baseline in the following section.

Shaw et al. (1998) have considered the sensitivity of Canadian coasts to a sea level rise of 0.65 meters. Their "coastal vulnerability index" combined data on seven variables: (*i*) relief and (*ii*) vertical land movement, which were considered indicators of inundation risk; (*iii*) lithology and (*iv*) coastal landforms, which were associated with resistance to erosion; (*v*) rates of erosion, which were considered indicators of sensitivity to coastal processes; (*vi*) wave energy, which was related to capacity for erosion; and (*vii*) tidal range, which was linked to both inundation and erosion hazards. A coast with a high sensitivity index would be in a region of low relief and unconsolidated sediments, high tidal range, high wave energy, and where sea level is already rapidly rising. Alternatively, a coast with a low sensitivity index would have high relief, a rocky shore with resistant non-eroding bedrock, falling sea level, low tidal range, and low wave energy.

Overall, the British Columbia coast has low sensitivity to sea level rise, due largely to a preponderance of high, rocky fjords and skerry shorelines. Due to steep topography, the area of land at risk of inundation is small relative to the other coastal provinces; yet, the majority of dwellings at risk in Canada are in British Columbia—about 8,900 to 18,700 by the 2050s—due to British Columbia's high coastal population density (National Round Table on the Environment and the Economy 2011: 69). This estimate may, however, be an order of magnitude low. Nationally, the annual cost of flooding of dwellings could be C$4 billion to C$17 billion by the 2050s (2011: 71). For British Columbia, the range could be from C$0.84 billion to C$13 billion dollars by the 2050s. Per capita costs of dwelling damages by the 2050s in British Columbia are expected to range from C$565 to C$2,146 per person per year.

The social and economic costs of sea level rise will not be evenly distributed. In terms of land area alone, for example, Canada's flood exposure due to climate change is minimal—just 0.14 percent of the nation's total land (Stanton et al. 2010: 8). However, what magnifies the economic impacts dramatically is that Canada's population and economic activity are concentrated in coastal areas, a fact that has significant consequences for British Columbia:

> The more pronounced this concentration is in a province or territory, the greater the impact it will feel—which is why British Columbia, with more than half its population concentrated in the coastal Vancouver and Victoria metropolitan areas, is particularly exposed, with up to $1.9 billion in annual damages from climate change-related flooding by the 2020s and up to $25.1 billion by the 2080s, depending on the scenarios used. That means 91 to 97 percent of total annual damages due to climate change would occur in British Columbia, even though it has only 0.8 to 1.4 percent of the land exposed to inundation due to climate change.
>
> (Stanton et al. 2010: 8)

Moreover, visible minorities and Indigenous peoples are over-represented among those exposed to sea level rise:

> In British Columbia, they account for more than 90 percent of the people living in areas exposed to inundation. Visible minorities make up 80 to 94 percent of the people exposed to climate-induced inundation overall; the vast majority of that population is in British Columbia, where they now make up 39 percent of the provincial population, and by the 2080s, this share is expected to grow to 49 percent. Aboriginal people make up 5 to 6 percent of the population exposed due to climate-induced inundation, with the greatest exposure in the 2020s in all scenarios.
>
> (Stanton et al. 2010: 8)

Low-income people will also be disproportionately affected (Stanton et al. 2010: 8). The average household income for the residents of areas exposed to inundation is lower than for the general population, ranging from 55 to 59 percent of the average for Canada as a whole. In British Columbia, the average income in areas exposed to inundation is 64 to 71 percent of the general-population average.

Salish Sea

In sum, Salish Sea landscapes most threatened by sea level rise include developed areas, soft-sediment bluffs and cliffs, and all low-lying areas, including shores, estuaries, and wetlands. Responses to sea level rise, especially armoring, may have a greater negative impact on coastal landscapes than the effects of sea level rise alone. In other words, doing nothing may be better than doing something. Regionally, sea level rise will cost hundreds of billions of dollars, and the socio-economic toll will not be equally shared across society. Visible minorities and Indigenous and low-income populations will be hardest hit.

Cultural Heritage Impacts

The physical impacts of sea level rise on archaeological heritage are generally well understood. Such study falls under the purview of environmental archaeology broadly (Dincauze 2000: 227–50) and coastal geoarchaeology specifically (Waters 1992: 249–90; Wells 2001: 149–82). The crucial lesson from geoarchaeology concerns that which keeps archaeological sites "whole"—that is, sediment. Sediment is the common matrix, fabric, or "glue" that keeps sites together, or "intact"; it is what gives sites "integrity" and what makes them "in situ" (Stein 2001; Wells 2001). From this perspective, sediment loss *is* site loss. As goes the coastal sediment (and soil), so goes the coastal archaeological site held together (or up) by that sediment—this rule applies equally to shell middens (Nutley 2005), medieval European churches (Murphy et al. 2009: 14), and historic American lighthouses (Erlandson 2012: 137).

Globally, and since the late 1990s, there has been a steady increase in the number and frequency of sea level rise threat studies (Cassar 2005; Cassar et al. 2006; Cazenave 2014; Chapman 2002; Fulford et al. 1997; International Council on Monuments and Sites [ICOMOS] 2006, 2010; Jones et al. 2008; Mitchell 2008; Perry 2011; Rick and Fitzpatrick 2012; Shearing 2007; The National Trust 2005; United Nations Educational, Scientific and Cultural Organization [UNESCO] 2007a, 2007b, 2008, 2009, 2010). Despite about a ten-year time lag (see, however, Easton 1990), this trend is now playing out in North America (Berenfeld 2008, 2015; Blankholm 2009; Caffrey and Beavers 2008, 2013; Erlandson 2012; Goetz 2010; Jones et al. 2008; Reeder et al. 2012; Westley et al. 2011).

Accounts from North American regions where such problems are already underway—the Arctic in particular (Blankholm 2009; Goetz 2010)—highlight the many complexities of the situation. This includes the speed at which erosion can occur, the extent of land that can be impacted in those time spans, and the potential for global and near-complete loss of low-elevation heritage landscapes. Pauline Goetz (2010: 12) describes the Arctic situation this way:

> Coastal erosion can sweep entire buried and built archaeological sites into the sea. It can also damage and weaken soil beds so that sites can collapse in upon themselves. In northern regions erosion thaws the permafrost causing slumping and the decay of organic artifacts. In the last decades various coastal areas are already seeing dramatic losses. In some cases, such as Barrow, Alaska, up to 30 meters of land is being swept away through erosion annually.

To understand sea level rise impacts on different kinds of archaeological heritage located in different kinds of environmental settings, it is useful to briefly consider the study of site-formation processes (Schiffer 1987). My logic here, simply put, is that if we want to know how archaeological sites "come undone," we should begin by understanding how they are "put together." Of specific concern here are natural transformations, or "N-Transforms," that operate to change an archaeological site's context. Analysis of natural site-formation processes is concerned with understanding the physical, chemical, and biological factors that affect site burial, alteration, and destruction (Waters 1992: 11). Such considerations have been made for the coastal zone, considered to be "one of the most dynamic landscape elements, with relative sea level, wave and tide energy, and sediment flux varying on time scales from seconds to millennia" (Wells 2001: 149).

While it is beyond the scope of the present inquiry to detail all the relevant aspects of coastal site formation processes, an overview of key issues is certainly needed. Those include (*i*) the factors that control the nature and transformation of the coastal zone, (*ii*) the processes that impact coastal archaeological sites (Table 2.1), and (*iii*) the factors that determine site survival through inundation (Table 2.2).

Three main factors control the nature and transformation of the coastal zone: sea level change, energy flux (waves, tides, currents), and sediment flux (Wells 2001: 149). Sea level rise impacts will be greatest on open or exposed coasts, low-lying coasts, and subsiding coasts. As illustrated in Figure 2.1, rising seas will impact a much greater area on low coasts than on steep coasts. Jon Erlandson (2012: 141) characterizes the land-loss situation in terms of coastal retreat:

> In decades to come, especially where cliffed coasts are present, the general principles outlined in Bruun's (1962, 1983) rule suggest that many shorelines will see horizontal erosion of 50 to 200 times any vertical rise, depending on the local geology, wave energy, and other factors.

Looking at Figure 2.1, and putting together some of the key concepts addressed so far, a sense of the scale of the sea level rise threat should be starting to form. Applying Bruun's rule to the sea level rise rates discussed earlier, low-lying, soft-sediment coasts could see horizontal erosion on the scale of 150 to 600 meters by 2300, as illustrated in Figure 2.2 (see, however, Cooper and Pilkey 2004).

My second area of focus relates to the processes that impact coastal archaeological sites under "normal" (i.e. non-sea level rise) conditions. This is important because sea level rise and population growth/impacts will be operating in

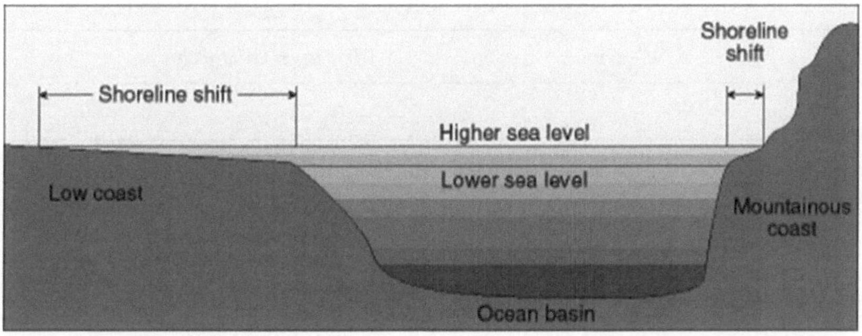

Figure 2.1 Impacts of sea level rise and resulting shoreline shift on low-lying coasts (left) and steep coasts (right).

Source: United States Geological Survey [USGS] (1997).

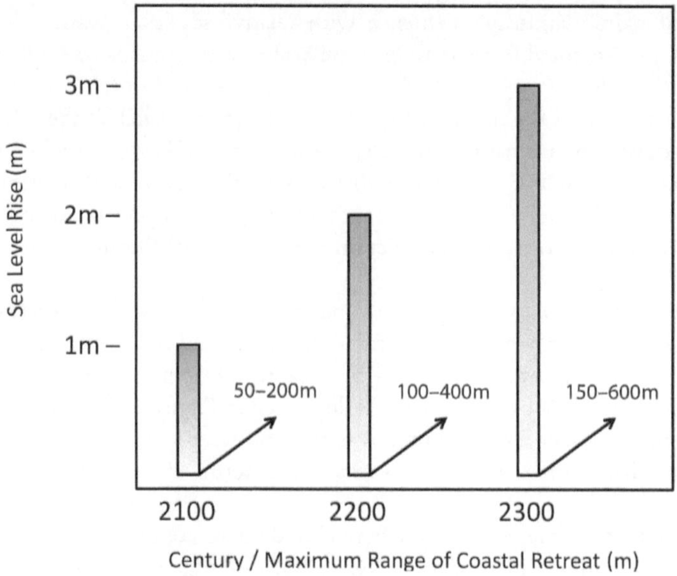

Figure 2.2 Global sea level rise estimates for the years 2100 (+1 m), 2200 (+2 m), and 2300 (+3 m), showing maximum range of shoreline shift erosion or coastal retreat.

Source: Data from BC (2013a), Horton et al. (2014), and Bruun (1962, 1983).

concert alongside and in addition to these "normal" processes. Those processes are outlined in Table 2.1, which links non-human processes with specific archaeological impacts. Of greatest concern here is how "marine processes" fit into the broader array of processes (left-hand column) and impacts (right-hand column).

Table 2.1 Major taphonomic and formation processes on California's Channel Islands

Process	*Description*	*Impact on archaeology*
Non-human processes		
Animal transporters	Deposition or removal of materials in a site by animals	Introduce non-cultural materials, remove some cultural materials
Argilliturbation	Shrinking and swelling of clay soils	Mixing of constituents
Eolian processes	Erosion and deposition of materials by wind, abrasion, and production of ventifacts	Deflation/destruction of sites, deposition of "new" sites

(Continued)

Table 2.1 Continued

Process	*Description*	*Impact on archaeology*
Faunalturbation	Burrowing and mixing of deposits by animals	Mixing, fragmentation, erosion
Floralturbation	Disturbance and mixing of deposits by plants	Mixing, fragmentation, erosion
Fluvial processes	Erosion and deposition of archaeological deposits by stream, creek, or other freshwater runoff	Destruction of sites, introduction of materials from other areas
Marine processes	Erosion and deposition of archaeological deposits by tidal surges and wave action	Destruction of sites, scouring of light fraction, introduce materials
Mass wasting/ gravity	Landslides, cliffing, etc.	Erosion, redeposition, and burial of archaeological record
Human processes		
Precontact human activities	Construction, cleaning, trampling, cooking, etc.; excavation of houses, storage areas, burial pits, etc.	Mixing, fragmentation, and destruction of archaeological record
Postcontact human activities	Building or road, construction, agriculture, bombing, looting, etc.	Movement, fragmentation, and destruction of archaeological record
Introduction of exotic animals	Overgrazing, stripping of vegetation and soils, trampling, burrowing, rooting, etc.	Fragmentation, mixing, deflation/ erosion

Source: Adapted from Rick et al. 2006: 571, Table 1.

By focusing on marine processes, in this case inundation, the complex nature of the sea level rise problem begins to take form, at least in terms of physical impacts on archaeological heritage. The specific issue of inundation in relation to Indigenous archaeological landscapes has been taken up by David Nutley, who summarizes the determinants of site survival through the inundation process, included here as Table 2.2 (Nutley 2005: 104–6; see also Tomalin 1998). As illustrated, consideration must be made for a variety of factors.

I have suggested that sea level rise cannot be fully understood without considering the contemporary landscape, both physical and social. I do so in the following section in the context of coastal sprawl, which provides an important frame for not just the sea level rise threat, but also the larger maritime heritage crisis.

Table 2.2 Determinants of archaeological site survival through inundation

	Determinants of site survival through inundation
Environmental factors	
Speed of water level rise	Rapid inundation minimizes the opportunity for mechanical, chemical, and biological attack. Slow or repeated inundation and exposure maximizes the degradation of sites and individual artifacts.
Fetch over water to land–water interface	A short fetch minimizes wave height and consequently minimizes wave-induced erosion.
Dominant wind direction	Sites facing the dominant wind direction are far more susceptible to wave action and destabilization from wind, rain, and falling trees.
Dominant wave direction	A direct response to the dominant wind direction and producing the most destructive of the on-shore forces influencing site survival.
Strength of water flow	Largely related to a riverine environment but also an aspect of estuaries and long-shore coastal currents. A strong flow of water will not only undermine sites but also result in greater dispersal of the component parts.
Tidal and non-tidal	Non-tidal water-bodies will result in less damage from frequent wetting and drying as well as minimizing the strength of water flow over the site. This may, however, be offset by the extreme changes between a saturated environment and a 'baked earth' environment during drought, as well as the impact of grazing animals on the exposed sediment flats during periods of low water.
Frequency and duration of inundation and exposure episodes	See "tidal and non-tidal" above.
Slope	A steep slope is accompanied by a marked increase in the gravitational factor. Earth, vegetation, rocks, and any associated deposits of cultural material are more likely to be destabilized and result in the loss of integrity of the original site formation.
Geological composition	Sites protected by coastal barriers, whether igneous rock, sandstone, or sand dunes, but with a healthy budget of sand or silt are much more likely to already be below sea level by the time that the coastal barrier is broken down and to therefore be spared the worst of the erosive forces of wave action.
Prograding/aggrading shorelines	Prograding/aggrading shorelines that provide habitation sites and artifact deposits with a protective depth of overburden may provide a sufficient buffer to ensure long-term survival during subsequent and relatively rapid inundation.

(Continued)

Table 2.2 Continued

	Determinants of site survival through inundation
Artifactual composition	
(i) Chemical impacts—susceptibility to permanent inundation in saline/fresh water	
Shell/bone	Shell and bone can survive extended periods in fresh and saline water.
Wood	Can survive extended periods in fresh and saline water.
Fibre (e.g. woven from leaf, grass, reeds)	Can survive extended periods in anaerobic freshwater and saline environments. In a marine environment, this is most likely to occur only in concretion formed through a combination of corroded iron and calcite from marine organisms.
Charcoal	Can survive almost indefinitely in marine and saline environments.
Stone	Can survive almost indefinitely in marine and saline environments.
(ii) Mechanical impacts	
(a) Susceptibility to abrasion	
Shell/bone	Highly susceptible to mechanical abrasion.
Wood	Highly susceptible to mechanical abrasion.
Fiber	Highly susceptible to mechanical abrasion.
Charcoal	Highly susceptible to mechanical abrasion.
Stone (flaked or ground stone artifacts and stone structures)	Moderately susceptible to mechanical abrasion, with greater resistance from metamorphic rock and some igneous rock and less resistance from sedimentary rock.
(b) Susceptibility to dislocation	
Shell/bone	Highly susceptible to dislocation.
Wood	Highly susceptible to dislocation.
Fiber	Highly susceptible to dislocation.
Charcoal	Highly susceptible to dislocation.
Stone (flaked or ground stone artifacts and stone structures)	Susceptibility to dislocation is proportional to size; small flaked stone artifacts moving sooner/easier than larger items.
(c) Susceptibility to biological attack	
Shell/bone	Mildly susceptible to biological attack.
Wood	Highly susceptible to biological attack unless buried.
Fiber	Highly susceptible to biological attack unless buried.
Charcoal	Minimally susceptible to biological attack.
Stone (flaked or ground stone artifacts and stone structures)	Mildly susceptible to biological attack through abrasion from grazing fish and sea urchins and other colonizing fauna and flora.

Source: Nutley 2005: 104–6.

Coastal Sprawl

> Our coasts are among the most rapidly growing and developed areas in the nation.
>
> (Crossett et al. 2004: 3)

Human intervention in the coastal zone or "anthropogenic factors" (Scott 2005: 254) are the leading cause of maritime heritage problems. A complete understanding of the sea level threat is predicated on a thorough accounting of the contemporary built environment, and the attitudes of the people associated with that environment. Towards that end, I consider here the role of population growth in maritime heritage destruction and sea level rise. As with the previous section, particular attention is paid to Washington State, British Columbia, and the Salish Sea region.

In my treatment of coastal sprawl, I address the issue of "amenity migration" and consider how that process can change communities' views towards heritage. While focus here is placed on growth, population decline or depopulation has its own set of implications for heritage stewardship (Ommer et al. 2007).

General Threats

Nearly a quarter of the world's population lives within 100 kilometers of the coast and less than 100 meters above sea level, resulting in population densities that are about three times higher than the global average:

> The attractiveness of the coast has resulted in disproportionately rapid expansion of economic activity, settlements, urban centers and tourist resorts. Migration of people to coastal regions is common in both developed and developing nations. Sixty percent of the world's 39 metropolises with a population of over 5 million are located within 100 km of the coast, including 12 of the world's 16 cities with populations greater than 10 million. Rapid urbanization has many consequences: for example, enlargement of natural coastal inlets and dredging of waterways for navigation, port facilities, and pipelines exacerbate saltwater intrusion into surface and ground waters.
>
> (Nicholls et al. 2007: 319)

Coastal areas are major destinations for tourism, which represents the fastest growing sector of the global economy (Creel 2003: 1). Migration into rural coastal communities by former urbanites escaping "city life" adds additional pressure (Gurran et al. 2007).

In the United States, the coastal population has grown steadily in recent decades, increasing from 47 million people in 1960 to 87 million people in 2008 (Wilson and Fischetti 2010). During this period, the Pacific coastline region gained the largest number of people (17 million). The Pacific region is the second most populated coastal region in the United States, and all five states that comprise it have the majority of their populations in coastal counties. Since 1960, the national coastal population has consistently added between 5 million and 10 million people each decade.

In Canada, between 1986 and 2001, the Canadian population increased from 25.3 million to 30.0 million people at a rate of 1.24 percent per year (Manson 2005: 4). Similarly, the population within 20 kilometers of a coast increased from 9.6 million to 11.5 million people at a rate of 1.32 percent per year.

In 2001, 38.3 percent of Canadians lived within 20 kilometers of a marine or Great Lakes shoreline, with the majority of the coastal population living within 5 kilometers of a shoreline (Manson 2005: 10); 23 percent lived on the Pacific coast (i.e. in British Columbia). Inhabited coastal areas represent 2.6 percent of Canada's total land area. Population tends to exhibit an exponential fall-off away from the coast, a point illustrated most dramatically in British Columbia, where steep mountain ranges constrain populations to narrow, low-elevation bands, as well as highly coastal-dependent economies (Manson 2005: 6).

Relative to the rest of the world, Canada shows similar trends in increasing total population count and density, particularly within the 20 kilometer coastal range (Manson 2005: 8). Manson concludes that through increasing population and densification, "the vulnerability of Canada's coastal population to hazards and climate change is increasing." Over the next 20 years, the coastal population is projected to grow by a million people, "increasing pressure on the environment through land-use changes and water demand, discharge of sewage and other waste, and emissions of pollutants" (BC 2006: vii). These problems are compounded by amenity migration.

In their discussion of governance responses to rapid coastal population growth, Nicole Gurran et al. (2007: 445) describe how "[t]he twin forces of rising affluence and population are altering coastal communities around the world. High amenity, environmentally sensitive areas—particularly attractive, non-metropolitan coastal environments—are witnessing a tidal wave of in migration from former urbanites." Amenity-rich areas like British Columbia's Sunshine Coast, discussed in Chapter Four, are struggling to accommodate growing numbers of people in areas with governance structures and physical infrastructure designed for occasional tourists:

> Places likely to attract amenity migrants commonly have high natural and rural scenery values, recreational amenities, a mild climate, and affordable residential land. ... For some migrants, particularly those who need to retain job connections to a major center, proximity to metropolitan areas remains an important criteria. ... Many migration decisions represent a culmination of a process starting with frequent holidays to a specific place, followed by acquisition of a second home, and finally permanent settlement. Smaller holiday towns are increasingly affected by amenity migration.
>
> (Gurran et al. 2007: 446)

Further, rapid population growth impacts the physical character of a community (Gurran et al. 2007: 452). In response to these changes, many coastal communities report a "loss of 'sense of place,'" as well as a "loss of sense of community or connection to social networks, because of the influx of new residents and visitors" (2007: 452). While the socioenvironmental impacts of amenity migration are now being explored in greater detail (Boucquey et al. 2012; Safford and Hamilton 2012; see also *GeoJournal* Vol. 76, No. 4, 2010), its impact on Indigenous maritime heritage landscapes remains much less well understood. As demonstrated in the *shíshálh* Coast study in Chapter Four, amenity migration is especially relevant to understanding coastal heritage destruction and degradation in the highly but unevenly populated Salish Sea basin.

Washington State

In 2003, approximately 153 million people in the United States (53 percent of the nation's population) lived in coastal counties, an increase of 33 million people since 1980. Ten of the 15 most populous cities in the United States are located in coastal counties (Crossett et al. 2004: 1). Most of the nation's most densely populated areas are located along the coast, with 23 of the 25 most densely populated counties in the United States being coastal (2004: 7). The Pacific region, which includes the coastlines of California, Oregon, Washington, Alaska, and the entire state of Hawaii, is the second most populated coastal region in the United States.

During peak vacation periods, coastal areas are also subject to major population influxes, and with more people "comes the need for increased infrastructure, which may lead to even more negative effects on resources" (Crossett et al. 2004: 1). Kristen Crossett et al. predict that over the coming decades coastal areas will also see a "growing proportion of older Americans and an unprecedented number of Americans reaching retirement age. This also has the potential to place demands on coastal resources as there will be more time for people to enjoy the many coastal amenities" (2004: 1). Median household income for coastal counties is approximately 17 percent higher than non-coastal counties (2004: 12).

Washington's population doubled over the last 40 years and is expected to reach 8.4 million by 2030, an increase of approximately 1.6 million persons from 2010. Net migration is expected to contribute 932,200 persons, or 57 percent, to the population increase (State of Washington Office of Financial Management 2010; see also United States Census Bureau 2005). Nearly 80 percent of Washington's population resides in coastal watersheds, most within the Puget Basin, which contains the City of Seattle, the economic center of Washington State, and the City of Olympia, the state capital.

Shoreline and nearshore landscapes have been—and will continue to be—severely impacted by population growth and development. The high level of human activity in the Puget basin "both partly springs from and leads to extensive use of nearshore ecosystems. Access to shipping, fishing and other commercial and recreational endeavors makes the region an attractive location for human settlement" (PSP 2011: 462). Coastal zone modification results in a wide range of changes in nearshore ecosystems, and these changes impact the shoreline, adjacent upland, and freshwater systems, and the Puget Sound estuary:

> Collectively, nearshore modification has resulted in shortening and simplification of shoreline over the past 150 years, from both direct (e.g., artificial structures) and indirect (e.g., disruption of shoreform sediment transport processes) modifications; the Sound has experienced a loss of over 1000 km of natural shoreline and the introduction of almost 400 km of artificial shoreline. ... This loss of convoluted shoreline has resulted in an overall loss of nearshore area, leading to disruption or loss of important ecosystem functions such as sediment, detritus and nutrient transport, loss of habitat and changes in species composition.
>
> (PSP 2011: 463)

Residential development of coastal bluffs is also a serious problem. Pressure to develop bluffs is "rising rapidly with the increasing population growth and urbanization of the Puget Sound region" (Shipman 2004: 89). According to Hugh Shipman (2004: 89), the demand for waterfront and bluff property is "driven primarily by access to the water and unimpeded views of the Sound and nearby mountains."

The physical impacts of development on Puget Sound coastlines are generally well understood (Huppert et al. 2009; Johannessen and MacLennan 2007; Myers 2010; Quinn 2010; Shipman 2004, 2008, 2010; Shipman et al. 2010a, 2010b). Despite such knowledge, a "chronic problem is society's tendency to ignore the fact that beaches and bluffs are not static systems, and that interfering with their dynamic processes may have undesirable consequences" (Johannessen and MacLennan 2007: 1).

British Columbia

Estimates of the precontact First Nation population of British Columbia range from 60,000 to 200,000 (BC 2007: 2). European diseases and other changes brought by early explorers decimated Indigenous communities, and by the 1850s their population had dropped drastically. As First Nation populations declined, European settlement along the coast continued to grow. As shown in Figure 2.3, in 1851 the population of British Columbia stood at 55,000; by 1951 it had grown to more than one million, and today it stands at over 4.5 million. In 2001, the Indigenous population was 170,025 and accounted for 4.4 percent of the total province.

In 2001, 76 percent of British Columbia's population, or just over 3 million people, lived in coastal regions, primarily in the Lower Mainland area around the City of Vancouver, the province's economic center, and in the southern and eastern parts of Vancouver Island, including the City of Victoria, the province's capital (BC 2007: 2). By 2036, British Columbia's population is forecast to

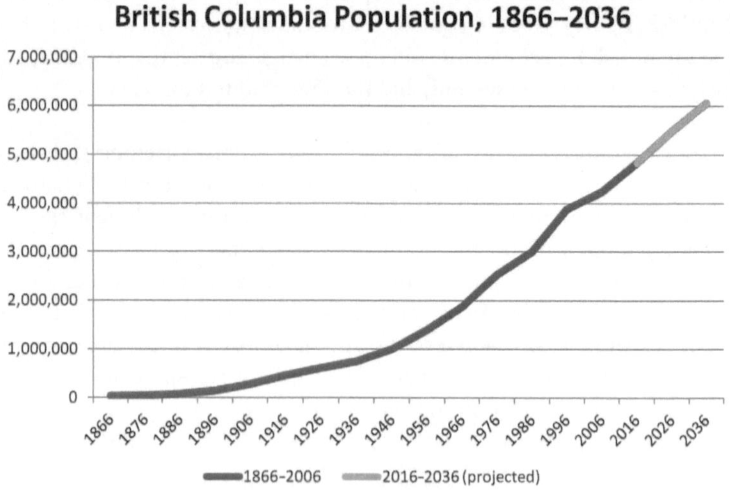

Figure 2.3 British Columbia population, 1866–2036.

Source: Data from BC Stats (2011a, 2011b).

exceed 1.6 million, an increase of about 36 percent since 2010 (BC Stats 2011a: 2; BC Stats 2011b: 1). The major factor driving population growth over this period will be migration: international migration will account for 77.4 percent of the population gain, followed by interprovincial migration at 17.3 percent, and natural increase (births minus deaths) at about 5.3 percent (BC Stats 2011a: 2). Future growth will be concentrated around already densely populated areas, specifically in the southern Strait of Georgia, around Vancouver, Victoria, and Nanaimo.

According to BC Stats (2011b: 4), the retiring baby boom cohort is "expected to lead to larger net inflows," resulting in population growth in amenity-rich areas, "particularly those outside of large urban cores" (e.g. Vancouver, Victoria). By 2036, the Vancouver Island/Coast region of southwest British Columbia, which includes the *shíshálh* Coast and the vast majority of the Georgia Basin, is projected to have almost five elderly dependents for every ten people of working age.

Intertidal and upper-shore zones in the Georgia basin are "particularly vulnerable to development because people are attracted by the views and easy access to beaches and the ocean" (BC 2006: 20). In 2001, an estimated 23 percent (60 kilometers) of the estuarine and other nearshore habitat in the Georgia Basin had been urbanized and the shorelines altered by dykes, seawalls, docks, and other uses.

Salish Sea

The Salish Sea was home to 4.1 million people in 1976; the region grew to 7 million people by 2001 and it is projected to grow to 9.4 million people by 2025 (GBPSEI 2006). Population growth in the Salish Sea basin is considered a "major underlying force contributing to the cumulative stresses on the land, air, water and other ecosystem resources. This growth presents significant environmental, economic and social challenges both locally and across the basin ecosystem" (GBPSEI 2002: 1).

People are attracted to the Salish Sea basin for its "spectacular natural amenities and quality of life, as well as its economic opportunities" (GBPSEI 2002: 2). As a consequence, a major contributor of population growth is migration from outside the basin: in British Columbia, the net immigration rate during the mid 1990s was about three-quarters of the total population growth rate; in Puget Sound Counties, almost two thirds of the growth over the 1990s was due to net immigration. Differences in growth rates suggest a shifting population from the urban to peripheral due to "perceived quality of life differences" and "expanding retirement population" (GBPSEI 2002: 2–3). Such growth affects all aspects of sustainability (environmental, social, and economic) at various scales (locally, regionally, and globally):

> Additional people generally require more jobs, goods and services, land for housing, businesses and transportation facilities, recreational space and associated environmental resources. Unless these activities are significantly modified to reduce their current impacts, they will continue to increase the stresses on the land, biota and other natural resources, as well as the human systems that support modern life.
>
> (GBPSEI 2002: 2–3)

Cultural Heritage Impacts

> Coastal archaeological resources around the world often coincide with dense contemporary human populations and a rapidly changing physical environment.
>
> (Reeder et al. 2012: 187)

Coastal sprawl can impact heritage landscapes in two ways: (*i*) directly, by altering coastal landscapes physically through development, and (*ii*) indirectly, by altering how coastal communities value, and thus define, view, and interact with, maritime heritage. Direct impacts often take place at or near the coastline; they may also be significant distances inland, as is the case with hydroelectric dams (Chen 2005). Such "interventions" include: aggregate extraction (including dredging); aquaculture facilities; docks and piers; energy production (e.g. oil, wind); gas mining (relative sea level rise induced by land subsidence); harbor infrastructure and activities (including navigational dredging); land reclamation, nearshore or offshore (e.g. wind farms); maritime navigation (e.g. ship and boat-induced waves); pollution; recreation and tourism; industrial resource extraction (e.g. fishing, logging, mining); river water regulation works (particularly dams); shore protection structures (e.g. breakwaters, groins, seawalls); seafront roads; tidal power facilities; and waterfront development, both industrial and residential (adapted from Eurosion 2004: 8–10).

Indirect human impacts are typically more complicated and subtle than direct ones, and there are innumerable ways in which human activities impact the behavior of coastal environments (Davis 2005: 533). These range "from the obvious situations where jetties at an inlet may impact beach erosion kilometers down the beach to more subtle situations where activities in an estuary can influence open coast morphodynamics" (2005: 535).

Little published quantitative data exists about the impacts of coastal sprawl on Pacific Northwest Coast archaeological heritage. Only highly contentious cases of site destruction seem to get any real coverage (Hutchings 2010, 2011, 2012a; Mapes 2009; Stapp and Longenecker 2009), either in the media or through professional publications. What typically goes undocumented, however, is the daily destruction, permitted or unpermitted, of the maritime heritage landscapes. By all appearances, the quantification of coastal site destruction and degradation is a low priority for archaeologists.

One exception is provided by Madonna Moss and Jon Erlandson (2008), who quantified cultural and environmental impacts on archaeological sites located in coastal Oregon State Park lands. Their study shows that "very few sites" are not actively eroding, this is due to seacliff retreat, riverine erosion, slopewash, trampling, wind erosion, or a combination thereof (2008: 34). Many sites were damaged and remain threatened by the construction of roads, parking areas, restrooms, picnic areas, and other park facilities, as well as by coastal erosion, weathering, sedimentation, dredging, clam digging, shipping, logging, oyster farming, and other ground-disturbing activities and processes. Littoral sites are considered especially vulnerable to looting because of their high visibility and accessibility.

Moss and Erlandson (2008) point out how concern for the "rapid destruction" of Oregon coast archaeological sites goes back more than 70 years. In 1935, for example, one individual noted that "erosion, agriculture, construction, and looting were heavily impacting coastal sites" (Moss and Erlandson 2008: 34). Nearly 60 years ago, Lloyd Collins (1953: 55, in Moss and Erlandson 2008: 34) observed:

> I haven't presented a bright picture of the situation, but one that spells doom for the prehistory of the coastal region of Oregon. ... In a few years the site count and excavatable sites will drop 50 percent … I recommend that excavation procedures be instituted immediately.

Writing about site destruction in Western Canada, Brian Spurling (1986: 465) had this ominous warning:

> The next generation of Western Canadian archaeologists … could be the last to have access to a reasonably intact [i.e. physically intact] archaeological record. ... To them will be the responsibility of maximizing returns

> from the remaining archaeological resource base. They will have the privilege of studying the few drainage systems not yet impounded for export, energy or irrigation. And they will exercise considerable influence over choices to excavate or preserve the rare, intact, multi-component sites not yet destroyed by urban and rural developments, transportation systems, or coastal erosion.

A different way to think about coastal development is in terms of social, cultural, and economic impacts. This includes what Rosemary Ommer et al. (2007: 3) refer to as *coasts under stress*, where "[c]hanges in natural environments have interacted with political, industrial, and social change to adversely affect the health of the people who live there, their communities, and the natural environments in which they are embedded." The lesson here is that coastal change not only involves the transformation or "wearing-away" of physical landscapes, but also the cultural attachments to those landscapes. Coastal change discourse, therefore, must be broadened to include the social impacts of such processes as economic change (Ommer et al. 2007), migration (Gurran et al. 2007), and recreation and tourism (Goodhead and Johnson 1996), among others. In this light, the discussion is not just about physical erosion, but also the erosion of community identity (French 2005), health (Dolan and Ommer 2008; Ommer et al. 2007), and heritage.

Summary

Coastal sprawl since 1950 has had a devastating impact on Indigenous heritage landscapes in North America (especially in resource- and amenity-rich areas that regularly draw developers, tourists, and retirees from nearby cities), and sea level rise over the coming centuries will destroy most of what remains. In this regard, late modern coastal change constitutes a crisis for Indigenous heritage, thus for Indigenous people. While coastal sprawl threatens the entire Salish Sea basin in a relatively uniform manner (along the core–periphery gradient), sea level rise poses a more complicated threat.

Due to the high occurrence of soft-sediment, bluffed, and low-lying coasts in the southern Salish Sea (Washington), that area is much more vulnerable to sea level rise than the steeper, hard-rock coasts that dominate the heavily fjorded northern basin (British Columbia). Yet, in the northern Salish Sea, steep topography means much of the population and infrastructure is closer to the shoreline, and rare low-lying coasts are that much more utilized and valued. This means that coastal heritage landscapes are under tremendous pressure in all areas of the Salish Sea, regardless of regional patterns. How the heritage industry is responding to this pressure is the focus of the next chapter.

3 Cultural Resource Management

> For thousands of years prior to European invasion, Indigenous peoples of the Western Hemisphere effectively managed their cultural and natural resources. Ancestral sites, sacred sites, community sites, and burial areas were cared for, utilized for ceremonies, and often discussed in oral traditions. After European invasion, American Indian, First Nations, and Indigenous peoples were disposed of their lands, and often denied access to areas of cultural and spiritual significance. Indigenous people in many areas often have to request permission to access their own sacred areas. As settlers moved into Indian country, looting and destruction of cultural, sacred, and burial sites became common, and millions of cultural items and thousands of human remains were removed from ancestral sites and obtained by museums, educational institutions, and collectors.
>
> (Steeves 2015: 124)

In Chapter Two, I discussed how the twin problems of sea level rise and coastal sprawl create the maritime heritage crisis. In this chapter, I discuss how resource management is the mainstream response to that crisis. I put resource management in historical perspective, focusing on the field/institution that is cultural resource management (CRM) with an emphasis on British Columbia. My account raises serious questions about state-sanctioned resource management, particularly as it pertains to Indigenous communities and their heritage. This lays the foundation for Chapter Four, where I examine the maritime heritage crisis as it is unfolding on British Columbia's "Sunshine Coast."

Resource Management

Melody Hessing and Michael Howlett (1999: 47–70) identify four stages of Canadian heritage stewardship, an abridged version of which is presented in Boxes 3.1 and 3.2. Their sequence reflects changes in the "institutions, processes and actors in the resource and environment policy sector" and shows government policy moving from (*i*) inaction to (*ii*) a concern with resource rents, then to (*iii*) a concern with conservation, and finally to (*iv*) the current resource management regime (1999: 47–8). We live in the "management era," dating from 1950 to the present (Box 3.2).

Box 3.1 The evolution of Canadian heritage policy, part I

Hessing and Howlett's first three stages are "inaction," "revenue," and "conservation" (1999: 48–9):

The Era of Government Inaction—pre-1800

In the colonies of Great Britain in the northern half of North America, it took several decades for governments to realize that they should take action on behalf of natural resources. Although some key resources of great financial or military value (such as gold or tall trees suitable for masts of wooden sailing ships) were reserved for the use of the Crown, other resources were simply used by settlers for their own purposes.

The first actions taken by governments involved their securing different forms of property rights, which ensured early mining and forestry operations that they would have a secure supply and the exclusive right to remove the resources located on the lands that they owned or leased. While land in some parts of the country (notably the Maritimes and parts of Quebec) was sold to private owners, elsewhere the overwhelming majority of land was never "alienated" but retained as Crown land. Resources located on private land ... required little regulation save surveying and title registration. ...

Once the demand for timber, mineral, or other resource "concessions" grew, it was possible for governments to begin to charge different kinds of rents for the use of their lands and resources, and Canadian government policy entered a new era.

The Era of Revenue Generation—1800–1880

The second period lasted for almost 100 years as governments in Canada developed a host of legislation containing a bewildering variety of charges for resource use. They included annual rents for mineral claims or timber limits, the development of extensive royalty or stumpage charges for the extraction of resources, and the development of embryonic taxation schemes levied against company profits. It was during this era the first government departments were established to deal with resources.

These "lands and forest" departments actively abetted the pillaging of Canadian resources by issuing many and cheap licenses to companies willing to extract resources at a rapid rate. By Confederation [1867], when most lands and resources were retained by the provinces, royalties from their sale by far amounted to the largest proportion of provincial government revenues. This great revenue made control over resources a subject of much inter-governmental competition. Between 1867 and 1900, the federal government attempted to use its powers over trade, commerce, and fisheries and other means to extend its control over provincial resources.

The Conservation Era—1880–1950
By the 1880s, the negative aspects of the policies of revenue enhancement were obvious in the deforestation of much of central and eastern Canada. In addition, due to technological improvements … governments were increasingly called upon to moderate the rate of resource extraction and to secure supplies for industries and towns over the long term and not simply on an annual basis. These calls led to the development of new administrative arrangements embodied in regulatory agencies … and long-term tenures. It was also during this era that Canada's system of national parks and forest reserves was created. … Prior to World War II, environmental functions, including resource management … were allocated to different ministries, legislated and administered largely at the provincial level. Administrative provisions for environmental protection were minimal, fragmented, and framed primarily in terms of human economic and organizational requirements.

Box 3.2 The evolution of Canadian heritage policy, part II

Hessing and Howlett's fourth and final stage is the "management era" (1999: 50–2):

The Management Era—1950–present
After the Second World War, federal reconstruction efforts targeted expansion of the national economy though resource development. … Throughout this period, efforts were also made at the provincial level to secure long-term resource conservation. This initiative involved long-term land zoning—such as the creation of forest and mineral reserves—as well as the establishment of new boards and agencies to secure resource supplies.

Over this period, the consolidation and force of environmental activity, in both legislation and administrative character, accelerated dramatically. The administrative structure for Canadian environmental protection was the focus of considerable attention. These changes reflected two major developments: an overall increase in the role and powers of the state in regulatory processes, and a corresponding increase in the complexity of legislation and administrative structures to facilitate this movement. In this period, environmental protection essentially underwent a transformation from de facto self-regulation by industry to more sophisticated state supervision of the regulatory process. … Regulatory models were adapted from many countries, especially the United States, where a flurry of postwar regulatory legislation was directed not only to pollution but also wilderness and species preservation.

> The establishment of ministries of the environment at both federal and provincial levels represents the formal recognition of an environmental agenda that both transcends and ... appropriates the responsibilities of existing resource departments. At the federal level, Environment Canada, whose jurisdiction included both resource management and pollution prevention, was created in 1971.
>
> Departments of the environment were created at the provincial level throughout the 1970s and 1980s, although they provided primarily environmental regulatory, rather than resource management, functions. The technique of the environmental impact assessment was introduced. Provincial legislation and administration was significant in its extension of regulatory provisions and its increased "get tough" policy, with greater punitive sanctions available for enforcement. Yet the legacy of structural fragmentation in jurisdiction and regulatory standards continues to make resource and environmental administration a complex and difficult process.

The 1949 British Columbia provincial election provides a useful point of departure for discussing the historical context of resource management, especially in light of the Great Acceleration (Steffen et al. 2015). In that election, William Andrew Cecil Bennett was returned to the British Columbia Legislative Assembly. While Bennett is known for many things, two stand out in particular: the hydroelectric dam on the Peace River that bears his name and his claim that "The finest music in the land is ringing of cash registers" (Bennett, in Mitchell 1983: 255).

In many ways, W. A. C. Bennett embodies what historian David Mitchell terms "The Rise of British Columbia." As Mitchell describes it, Bennett came to power when the province "was ripe for development, a rich frontier ready for experimentation and innovation" (1983: 256). During the 1950s, the Western industrial world, especially British Columbia:

> embarked upon a period of economic expansion and material prosperity that, in retrospect, we realize was exceptional. That decade in particular has become perceived as a time of social, political and economic innocence which later generations both long for and deride.
>
> (1983: 256–7)

It is in this light that resource management came about in British Columbia, supported in large part by the passage of environmental protection legislation that in many ways defines the "environmental movement" of the late 1960s to early 1980s. At the time, Bruce Mitchell and W. R. D. Sewell (1981: 2–19) identified these nine barriers to "effective" coastal resource management in Canada:

1. Incomplete information: too little data about resources;
2. Divided and fragmented responsibility: ambiguities and inconsistencies in jurisdiction over resources;

3 Poorly integrated institutional arrangements: as a consequence of point two, realizing *integrated* management of resources is difficult;
4 Trade-offs and compromises: resources may be traded-off to obtain other things, or used as mechanisms to achieve gains in other areas;
5 Diffused and changing public interests: the resource management policy field experiences varied and changing public interest;
6 Educational structures: organizational structures at universities which are based on disciplines provide little opportunity for students to be exposed to the necessary cross-section of concepts;
7 External events and decision: resource management is influenced by events outside the country, i.e. globalization;
8 Lack of control: foreign investment reduces domestic control over resources, i.e. economic globalization; and
9 Political commitment: political leaders are often seen as having low commitment to environmental issues.

Hessing and Howlett (1999: 232) summarize the history of resource management in Canada as follows:

> Over a long period, Canadian policy can be understood to have shifted from a system of laissez-faire resource exploitation to a system in which governments are heavily involved in a complex regime of resource management and environmental protection. The shift encompasses several features. Canada's history as a staples economy has encouraged resource extraction, traditionally driven by private interests. The growth of government's role in resource extraction parallels the general expansion of the Keynesian welfare state, increased public knowledge about the potential adverse effects of resource degradation, and increased competition and conflict over access to resources.

How CRM fits into this history, especially given its close ties to natural resource management, is addressed below.

Cultural Resource Management (CRM)

> [T]he granting of permission to build factories or other structures at places where [Indigenous heritage] sites are located … should be made contingent on the provision by the interested parties of funds for the investigation of such sites before construction commences. We cannot prevent urban expansion and industrial development, but by intelligent legislation they could be turned from a bane to a boon to archaeology.
>
> (Borden 1950: n.p.)

According to Eldon Yellowhorn (pers. comm. 2014), the notion of a "cultural resource" emerged in the 1950s (Meighan and Eberhart 1953), modified soon thereafter in the 1960s (Wendorf 1963), at which point the concept became linked with free enterprise. For Yellowhorn:

> The resource paradigm was a good fit too for public government who saw no difference between natural and cultural resources. In Canada, the constitutional division of powers meant that all archaeological resources are the property of the province in which they are found.
>
> (2014: n.p.)

Indeed, Thomas King (1998: 6–7) posits the term CRM was invented by archaeologists "to equate what they did with natural resource management." While one might expect the term to mean the management of cultural resources, it is commonly used, primarily by archaeologists, in a much narrower sense to refer to "managing historic places of archeological, architectural, and historical interest and to considering such places in compliance with environmental and historic preservation laws" (1998: 7):

> Back in the 1970s, archaeologists in the United States faced a challenge. New laws had been enacted promoting the protection of "natural resources" on the one hand and "historic properties" on the other. Government agencies were being required to conduct environmental impact assessments of their actions, seeking ways to protect the environment. How could archaeologists be sure that the places they were concerned about—archaeological sites—were protected by these legal requirements?
>
> (King 2008: n.p.)

Below, I describe archaeology/CRM as the mainstream response to coastal change, including sea level rise and coastal sprawl. The response is shown to be fundamentally about economics, with science and scientific discovery as the practice's moral center. Approaches are shown to vary depending upon whose heritage is at risk and who is doing the evaluation.

CRM and Sea Level Rise

> It was in the context of the irrevocable and dramatic loss of coastal sites that one particular conservation theme arose: "We'll never save everything, so hard decisions are needed as to which to 'let go.'" Coastal loss also raised the issue of heritage input into broader planning: "…I presume that major sea level rises will be a non-heritage issue (i.e., of such wider concern that it will be dealt with by other authorities)."
>
> (Cassar 2005: 3)

Having already gained a general understanding of sea level rise and its impacts on tangible heritage, I now turn attention to how people are responding to the situation. My goals here are to (*i*) briefly survey the variety of responses and (*ii*) identify the perimeter of response discourse as it stands today.

It is clear from the literature (see list below) that when it comes to heritage landscapes, sea level rise is largely the realm of the state and its experts, notably archaeologists and resource managers. Consider, for example, the following introductory statements about the sea level rise threat. While "[m]itigation measures, including excavation and recording of sites before their loss, and relocation of buildings, have already been initiated in some places," in light of sea level rise, "where hundreds of historic assets are threatened by natural processes full mitigation is likely to be financially unrealistic. Heritage groups and agencies will need to prioritize their resources for sites and buildings of special interest" (Murphy et al. 2009: 11). Put another way, "As all sites cannot be protected, it is essential that heritage managers know which sites and landscapes are most at risk so they can prioritize resources and decision-making most effectively" (Westley et al. 2011: 351).

May Cassar's comments at the beginning of this section concerning which sites to "let go" and "broader planning" raise the twin issues that frame sea level rise responses as they pertain to archaeological heritage. These are (*i*) prioritization: what "values" inform decision-making (e.g. what to "save"); and (*ii*) mitigation: the attendant "approaches" that are pursued. Together, these elements comprise the key points of departure for a "response."

One way to survey the perimeter delineating sea level rise discourse is by looking at the core values that underpin, inform, and guide proposed responses. Such values appear frequently in the stated objectives and rationale for most responses. Below are a few examples with emphasis added:

> Because many countries lack sufficient laws to protect archaeological sites, particularly in coastal areas where development is extremely desirable, *archaeologists are working quickly to survey and record the wide array of prehistoric settlements in these regions before they are disturbed or destroyed.*
>
> (Rick and Fitzpatrick 2012: 135)

> Few, if any, other disciplines can so successfully marshal the evidence of human activity and that of palaeoenvironmental change or, as a matter of course, join forces with geographers, climatologists and other environmental scientists to understand humanity within a framework that is at once historical *and* ecological, and thus able to critique Western narratives of environmental change. Without doubt, then, *archaeologists can usefully contribute to academic, popular and political debates on the prediction and the management of climate change* (*cf* CBA 2007).
>
> (Mitchell 2008: 1096)

> We highlight the significance of coastal archaeological sites for helping better understand contemporary environmental and cultural issues, *underscoring the need to preserve or salvage these sites for their significant research value.*
>
> (Reeder et al. 2012: 187)

> [Global warming] presents potentially large problems for the future of site and landscape management. However, there are a number of ways that may be considered advantageous for addressing this threat. Primarily, *monitoring needs to be conducted* with regard to archaeological sites.
>
> (Chapman 2002: 244)

> We need a concerted, collaborative, and global effort to *bring the problem to the attention of government leaders and the general public.*
>
> (Erlandson 2012: 141)

> The conclusion is that although research and policy is moving ahead, much more needs to be done in order to *understand and preserve these sites before it is too late.*
>
> (Milner 2012: 223)

A core value expressed in these and other such assertions is that heritage is important for its "research value" and ability to produce scientific knowledge. A basic assumption is that the loss of the "archaeological record" is the most pressing issue when it comes to sea level rise. Recognizing that not all sites can be conserved, the approach suggested is to learn as much about the sites as possible before they are destroyed, and focus efforts on those deemed most "significant." Alongside a call for public education, the response from archaeologists is that more archaeological research is needed. This presumes that science is the limiting factor in the heritage crisis. The central role of government and CRM in overseeing the response is also largely taken for granted.

In raising the notion of "major sea level rise" being a "non-heritage" issue, Cassar (2005) is highlighting my second concern: What defines the perimeter of the mainstream response? When heritage specialists talk about sea level rise impacts and responses, which issues are "in" and which are "out"? Where are the walls of the silo to be placed? As Peter Mitchell (2008: 1093) asks:

> What, then, is the responsibility of the archaeological community and of the individual archaeologist to these challenges? Do either, in fact, even have a particular responsibility *qua* archaeologists, above and beyond those that they may have as concerned citizens and inhabitants of the planet? And, faced with the harsh realities that global heating will bring, is archaeology anything more than a dilettantish luxury?

Sea level rise responses are typically framed within a set of generic policy options. This limits responses to between three and five options, depending upon which set of options one uses (BC 2013a: 19–22; BC 2013b: 21–2; English Heritage 2006a; Neal et al. 2005). The effect is that options are constrained to two sets of decisions: (*i*) *action* (intervention) versus *inaction* (no intervention) and (*ii*) *advance*, *hold*, or *retreat*. The individual categories include (Fulford et al. 1997):

1 *Inaction*, or no active intervention, leading to inundation;
2 *Managed retreat*, or managed realignment, wherein a new landward line of defense is set;
3 *Hold the line*, wherein coastal defenses (seawalls) are built/modified to hold back the sea; and
4 *Advance the line*, or move seawards, whereby defenses are built seaward of current ones.

In Canada, strategy four, "advance the line" or "attack," is not considered a viable response to sea level rise (BC 2013b: 22). In British Columbia, then, responses to sea level rise are limited to inaction (do nothing), managed retreat (accommodate and withdraw), and hold the line (defend in place).

As discussed, there have been a number of formalized and comprehensive responses developed in the past decade to deal with coastal erosion and climate change more broadly (e.g. Berenfeld 2015; English Heritage 2007). In England, the agency charged with overseeing the management of the country's archaeological landscapes, English Heritage, has developed a comprehensive guidance program—the Rapid Coastal Zone Assessment Survey (RCZAS)—to "enhance the coastal National Monuments Record (NMR) and county Historic Environment Records (HER)" (English Heritage 2007: 12). The RCZAS process is laid out in the document "A Brief for English Heritage Rapid Coastal Zone Assessment Surveys" (English Heritage 2007). That document provides an outline methodology for "rapid baseline coastal surveys" commissioned by English Heritage. In addition to providing a "broad assessment of the likely archaeological potential and vulnerability of all stretches of the coast," the RCZAS is designed to provide a "curatorial response to individual applications for commercial developments or schemes," an enhanced record for "coastal heritage assets," and an "overview of coastal change from the Late Upper Paleolithic onwards" (2007: 3). Like the responses discussed above, state agencies and experts are at the center of the English Heritage response. Scientific discovery is a prime objective and the heritage list (or schedule) is the organizing framework. The term *heritage list* refers to the state's register of "official" sites, whereby listed sites are subject to state law and unlisted or "unauthorized" landscapes are not (Harrison 2013; Smith 2004, 2006).

Climate change and sea level rise responses proposed for heritage landscapes on United States National Park Service (NPS) land follow the RCZAS pattern insofar as they take the state for granted and make scientific discovery (i.e. "data recovery") the nexus of the response. In their discussion of coastal National Parks, Maria Caffrey and Rebecca Beavers (2008: 87) stress that "[t]ourism is a vital part of protecting cultural resources because tourist dollars can contribute to protecting sites of cultural significance." Taking an emergency "triage" approach to climate change, Michelle Berenfeld (2015: 8) writes:

> A 'triage' approach would … affect how NPS distributes limited resources. Every dollar that goes into maintaining the foundations of a fort that sits, for

> now, at sea level is one more dollar that could be spent on archival research, analytical documentation, artistic interpretation, and scholarship about the history of that place before the site is permanently flooded and made inaccessible. … The budget will always be finite, and time and money spent shoring up things that will only need to be shored up again and again until they are, in fifty or a hundred years, unsalvageable, will be time and money that could have been—but was not—spent on creating a future history of that doomed place.

According to Berenfeld (2015: 7), the NPS "must get into the business of creating an *afterlife* for cultural resources, even as it may continue to preserve their physical lives for a while longer."

In sum, the mainstream response to climate change and sea level rise threats is the government's response, which is CRM. The response consists of (*i*) data collection (e.g. surveying, mapping, listing), (*ii*) ranking those sites according to their value (valuation), typically their scientific value, and (*iii*) allocating resources (money) according to those sites deemed most "at-risk." Written by state-sanctioned heritage experts who derive their livelihood from the state, it is perhaps not surprising that neither state nor institution nor expert is ever conceived of as being part of the problem. Although economics underpins much of the decision-making process, science and scientific discovery are presented as the heart of the response. Also not considered is that for most places, no government funding exists to support "creating a future history" for the vast majority of heritage landscapes, especially Indigenous heritage landscapes. Unofficial sites not recognized or listed by the state are rarely considered in mainstream responses.

CRM and Coastal Sprawl

For archaeologists and cultural resource managers, the maritime heritage crisis means difficult decisions must be made as to what to "save" and what to "let go." Here, I consider two very different approaches to heritage management in Washington State. The first utilizes a "National Maritime Area" approach to save colonial maritime heritage from development. The second takes a "Risk Management" approach to help protect Indigenous maritime heritage from development. At issue here is who is speaking and whose values they are representing. This raises important questions about heritage experts and the nature of archaeology/CRM.

A National Maritime Heritage Area Approach

In 2005, King County and the City of Seattle jointly released the document *Seattle/King County Task Force on Maritime Heritage* (Task Force 2005). The report opens with this introduction by politician and Task Force Chair Jim Compton:

> Our region's history is tied to the sea and its industries. This is a rare moment when the "stars are aligned" to launch a long-overdue effort to preserve Puget Sound's rich legacy of maritime history. Centuries of maritime activity on Puget Sound have created an extraordinarily rich body of traditions, artifacts, and history.
>
> (Task Force 2005: 2)

In this document, maritime heritage is conceived of in very specific terms: European and Euro-American maritime technology, exploration, discovery, and industry (what the Task Force refers to as "working waterfronts"). Indigenous heritage is mentioned once in an appendix as a potential "theme."

The Task Force's objective, described in their Executive Summary, was to "[c]reate a vision and chart a course for a sustainable maritime heritage presence in Seattle, King County and Puget Sound." The objective, in a nutshell, is to "house," "brand," and "market" Seattle/King County's maritime heritage under a "single leadership" with the goal of delivering a "cohesive," "unified," and "authentic" message regarding maritime heritage. Significantly, there is no recognition of or discussion about climate change and sea level rise. There is also very little specific information provided regarding population growth and the impacts of development on heritage. Tourism is offered as a solution to at-risk maritime cultural heritage, and it is argued that obtaining a federal "National Heritage Area" designation would facilitate tourism.

In 2009, the Washington State Department of Archaeology and Historic Preservation (DAHP) produced the draft report *Washington State National Maritime Heritage Area Feasibility Study* (DAHP 2009), building on the 2005 Task Force call for a National Heritage Area. In contrast to the 2005 Task Force report, the 2009 document includes discussion of Indigenous history and contemporary interests in the Heritage Area. Indeed, the economic benefits for tribes overall is emphasized, since tourism and related economic development is offered as "one of the important potential benefits of designation as a National Heritage Area" (DAHP 2009: 11). Towards this, maritime commerce, past and present, is given central stage in the proposal, and heritage is viewed as a commodity and potential money-maker for willing communities. No critical discussion of the socioenvironmental impacts of maritime industry in Puget Sound is offered.

The final 2010 version of the project, *Washington State National Maritime Heritage Area Feasibility Study for Designation as a National Heritage Area*, like its predecessors, does not address any climate change issues for coastal heritage (DAHP 2010). Likewise, it does not address any impacts or management issues for cultural resources; indeed, the subject of archaeology is scarcely mentioned. As such, this report cannot be seen as being motivated by the protection, preservation, and/or conservation of heritage sites. Instead, the proposed project to establish a National Maritime Heritage Area is fundamentally about economic (re)development through tourism, under the guise of rejuvenating a local sense of maritime heritage.

A Risk Management Approach

While the National Maritime Heritage Area study constitutes a "tourism" response to the destruction of settler heritage, the Tse-whit-zen case discussed below provides an account of the impacts of development and archaeology/CRM on Indigenous heritage (Charles 2010; Hutchings 2011; Mapes 2009; Stapp and Longenecker 2009). The destruction of the Tse-whit-zen village site constitutes what archaeologists Darby Stapp and Julia Longenecker call an "archaeological disaster."

In 2002, the City of Port Angeles sold to the Washington State Department of Transportation (WSDOT) a large waterfront parcel to be used as a dry- or graving-dock. Construction and use of the graving dock, originally projected to cost around US$19 million, was a massive and unexpected infusion of wealth (and power) into the relatively small community of Port Angeles (Stapp and Longenecker 2009: 41); however, over a two-year period WSDOT would "disrupt 276 full burials and 500 partial burials of Lower Elwha Klallam ancestors. The costs for recovering the burials and the millions of artifacts were staggering." Indeed, the final price tag for the uncompleted graving dock was about US$85 million.

Tse-whit-zen represents a heritage disaster; however, it is by no means an isolated event (Stapp and Longenecker 2009), and sea level rise and continued population growth significantly complicate matters. The mainstream response to coastal change is encapsulated in what Stapp and Longenecker call the "risk management approach," which is a checklist and mitigation approach to heritage stewardship whereby site loss is taken for granted (Hutchings 2011). Lower Elwha Tribal Chairwoman Frances Charles questions the utility of such an approach: "Everyone would go back and say, 'What we would we do differently?' It's good to have a checklist. But what is the next crisis going to be?" (Mapes 2009: 117; see also Hutchings 2010).

While the state eventually abandoned the project, that action did not bring closure for the Lower Elwha Indigenous community, for "breaking ground at Tse-whit-zen uncovered not only the past of this place but its present" (Mapes 2009: 215):

> The true history under every footfall, along every shoreline, is often silenced by collective amnesia. But here, on a twenty-two-acre waterfront property, the ground spoke: of the Indian village here, and of a river that once sustained an entire ecosystem and way of life. Of uncounted Indian burials, ancestors of the Klallam people. Of the sawmill built right over their village and cemetery, transforming the look, but not the truth, of this ground.
>
> (Mapes 2009: xi)

The Tse-whit-zen story thus parallels the history of colonialism and Native–non-Native relations on the Pacific Northwest Coast, where they, for a variety of reasons, "don't understand each other's history" (Mapes 2009: 120). Frances

Charles blames the City of Port Angeles: "They knew what was here 150 years ago. They knew the heritage, of what was here. They can't sit there and be unaware. They ignored it because of their greed" (Mapes 2009: 117; see also Charles 2010).

CRM in British Columbia

The above examples from Washington State highlight the complex nature of archaeology and CRM. This institution seems driven by capitalism and scientific discovery. As such, mainstream responses appear to be more interested in promoting development than protecting heritage. As discussed below, the situation is no different in British Columbia.

The history of archaeology in British Columbia has been dealt with by various historians including Douglas Cole (1985), Robert West (1995), and Susan Roy (2010). Cole describes the early period of archaeology, arising out of the great "scramble" to collect artifacts and skeletons of Northwest Coast Indigenous peoples. West relates the "professionalization" and "scientification" of the discipline of archaeology in the province. He takes a critical approach to argue that this formalization in the 1950s by Charles Borden reflected how a small group of elite academics took control over Indigenous heritage. Building on this work, Roy brings this narrative into the present, suggesting the project of archaeology and the construction of "prehistory" represents the appropriation of Indigenous heritage, part of the larger colonial project of "forgetting" (see also Mapes 2009). While West traces the "rise" of academic archaeology in British Columbia (compare with Smith 2004), Brian Spurling (1986, 1988) describes the development of CRM, discussed in detail below.

Canada has no federal overarching heritage legislation. In British Columbia, cultural resources are defined in relation to the *Heritage Conservation Act* (HCA). As articulated by the province, the Act is designed to "encourage and facilitate the protection of conservation of heritage property in British Columbia" (BC 1996). Archaeological sites are managed by the British Columbia Archaeology Branch, currently run out of the Ministry of Forests, Lands and Natural Resource Operations. In this sense, cultural heritage is viewed as another resource alongside, for example, minerals and lumber. The HCA's most noteworthy feature is an 1846 cut-off date, a caveat that means British Columbia archaeology focuses almost exclusively on Indigenous heritage—thus living Indigenous people.

Under the HCA, British Columbia's Archaeology Branch is responsible for "maintaining and distributing archaeological information" and the permitting of development "within protected sites" (BC 2012: n.p.). The Archaeology Branch's role is described on their website as being "not to prohibit or impede land use and development, but rather to assist the development industry, the province, regional authorities, and municipalities in making decisions leading to rational land use and development" (BC 2012).

Archaeological work throughout the province is authorized through a permitting system and a provincial heritage registry is maintained of all known archaeological sites, heritage sites and objects, heritage wrecks, and other types of sites. The CRM process in British Columbia has been described as follows:

> Those seeking to develop an area that either contains a recorded archaeological site or has not been assessed for archaeological remains are responsible for the logistical and financial planning of this process. Depending on the scale of the proposed development, archaeological sites are subject to varying levels of assessments known as Archaeological Overview, Impact and Alteration. In this system, landowners hire qualified archaeologists to determine the levels of investigation required and undertake this work under provincially issued permits.
>
> (Archaeological Society of British Columbia 2011: 1)

This approach is commonly known as a "user-pay model," where the project proponent is responsible for paying for the archaeological permitting and mitigation process. Archaeological work conducted for pay, usually in the context of development, is referred to as "contract archaeology" (Stapp and Longenecker 2009) or "archaeology-for-profit" (Muckle 2013). In British Columbia, therefore, CRM is archaeology designed to facilitate the public's compliance with the HCA and Archaeology Branch policies. In this light, CRM may be construed as "compliance archaeology."

The British Columbia Archaeology Branch summarizes the site evaluation process as follows:

> When the benefits of a project outweigh the benefits of archaeological preservation, the branch will work with the proponent to determine how the project can go ahead with minimal archaeological resource loss. Where the loss of significant archaeological values cannot be avoided, the branch ensures that appropriate compensatory measures are implemented.
>
> (BC 2012: n.p.)

The concept of "significance" is thus critical. The Archaeological Impact Assessment Guidelines Appendix D Checklist of Criteria for Pre-Contact Site Evaluation identifies four features to evaluate significance: scientific, public, ethnic, and economic. The Appendix E Checklist of Criteria for Post-Contact Site adds to these the criteria of historic significance and integrity and condition. The two sets of evaluative criteria somewhat reflect the date of 1846 established by the HCA in relation to what sites and objects are automatically protected by law (Section 13.1.d)—in other words, whether it is settler or Indigenous heritage. However, it is unclear to what extent each category of significance is weighted—for example, whether a site with high "ethnic" significance but low "scientific," "public," and "economic" significance warrants increased conservation efforts.

In contrast to academic archaeology publications, comparatively little has been written about CRM in British Columbia. In an attempt to better understand the history of commercial archaeology here, Marina La Salle and I undertook a basic survey of the practice (La Salle and Hutchings 2012). Specifically, we sought to bridge recent political economy studies of CRM in other parts of Canada (Birch 2007; Ferris 2002; Zorzin 2010, 2011) with an older political economy analysis of western Canadian CRM (Spurling 1986).

In 1986, Brian Spurling completed an exhaustive study of Western Canadian archaeological resource management. In the process he found evidence of "serious problems" (1986: 464). Viewing CRM as a *policy science*, insofar as it represents "the legislated conservation and preservation activities of the archaeological profession" (1986: 11, 19), Spurling had this to say about the rise of commercial archaeology: "The transformation of archaeology into a business is a very recent phenomenon, one which occurred entirely within the last decade as a response to the passage of provincial heritage legislation" (1986: 291). Initially viewed with "trepidation and suspicion" by the archaeological establishment, commercial consulting "proved to be the only areas of disciplinary expansion through the later 1970s and early 1980s. As the ranks of universities, museums and government agencies were filled, students graduating with Masters and PhDs perforce gravitated towards private sector employment" (1986: 292).

Spurling, from his vantage in 1986, identified four "gaps and problems" for archaeology: (*i*) "academic archaeologists can still argue that [CRM] studies make few theoretical or methodological contributions to the discipline"; (*ii*) the interested public, including avocational archaeologists, are "mostly excluded from meaningful involvement"; (*iii*) the wider public remains basically unaware of the "results of and need for archaeological activity"; and (*iv*) archaeological heritage "is still being lost at uncontrolled and unknown rates" (1986: 464). Additionally, and perhaps most significant, archaeology "cannot be defended" against allegations of resourcism (1986: 500).

Building on this work, La Salle and I sought to characterize contemporary commercial archaeology in British Columbia (La Salle and Hutchings 2012). This included an historical focus on the permitting process in this province, and specifically how many permits have been issued annually since permits were first issued in 1960. The trend in annual permits issued has been one of significant growth which, we postulate, is directly connected to larger economic trends.

Indeed, most archaeologists work for multi- or trans-national consulting corporations, which are hired by development proponents to complete necessary impact assessments and prepare reports for the government (La Salle and Hutchings 2012; see King 2009 and Dent 2012). Analysis showed that 97 percent of archaeology undertaken in British Columbia in the year 2011 was in the form of compliance archaeology (Figure 3.1). This is a critical point for it confirms that mainstream archaeology is CRM, despite intellectual efforts to distance academic (research) archaeology from CRM (Dunnell 1984).

In this way, nearly all archaeology/CRM performed in the province is related to development (La Salle and Hutchings 2012). Indeed, some archaeologists have critiqued the Archaeology Branch for "facilitating 'demolition permits' rather than regulating any modern heritage conservation practice," suggesting

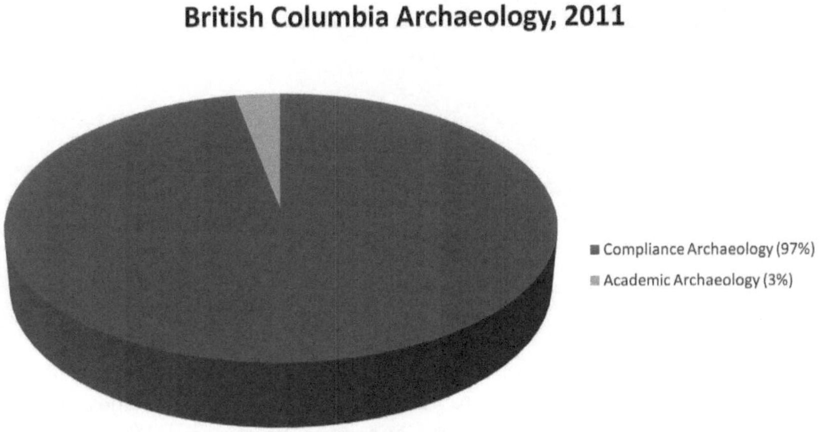

Figure 3.1 British Columbia archaeology defined in terms of how it is practiced. Virtually all archaeology is compliance archaeology, or cultural resource management, at least as measured in 2011 in British Columbia, Canada.

Source: Data from La Salle and Hutchings (2012: 10).

that "no archaeological site is safe from the wrecking ball in British Columbia" (McLay 2011: 5–7).

As a process governed by state policy and practiced within a commercial setting—both predicated on capitalist ideologies of growth, development, and progress—CRM, therefore, has built into it some fundamental limitations. Critically, there is little quantification about regional site loss, either in part or in total. Without this knowledge, there is no baseline to evaluate whether or not CRM has been a "success" or a "failure" at conserving archaeological sites as per the HCA.

Summary

The mainstream heritage response to the maritime heritage crisis is archaeology/CRM, an endeavor rationalized by archaeological science and rooted in economics. In relation to sea level rise, this has taken the form of a triage approach whereby heritage sites are evaluated based on risk and value. Where development is the primary threat, the response varies depending on the project and the nature of the heritage itself. Colonial maritime heritage is often treated differently, valued for its tourism potential and seen as building community. Conversely, Indigenous maritime heritage is frequently viewed as an impediment to economic development; archaeology/CRM in this context can result instead in the fragmentation of community. That CRM is directly linked to the development responsible for both sea level rise and coastal sprawl indicates a fundamental conflict of interest. As such, the question remains as to how effective archaeology/CRM is as a response to the maritime heritage crisis. Evaluating this question is the subject of the next chapter.

4 The *shíshálh* Coast Study

> We, the *shíshálh* Nation, openly and publicly declare that we have Aboriginal Title and Aboriginal Rights to our Territory, including the lands, waters, and resources that have been ours since time immemorial. We have been given the responsibility from the Creator to care for our Territory. Our Territory sustains our people, maintains our Indigenous way of life, and is integral to our identity as *shíshálh*. We have always governed ourselves and our Territory, and have never relinquished our authority or jurisdiction over such. We assert our collective right to live as a distinct people.
>
> (*shíshálh* Nation Declaration, Sechelt Indian Band [SIB] 2013: 1)

In this chapter, I consider the issue of coastal change in the context of *shíshálh* (Sechelt) First Nation maritime heritage. *shíshálh* traditional territory, and thus history, is within the region commonly known as "The Sunshine Coast," located about 10 nautical miles (20 kilometers) northwest of Vancouver, British Columbia (Figure 4.1). The main objective of the study was to see what coastal change impact looks like "in practice" and "on the ground" with a view to assessing the effectiveness of cultural resource management (CRM) at protecting archaeological sites.

I designed the study to capture a cross-section of landscapes and issues as they exist along a mostly uninterrupted length of *shíshálh* coastline. The *shíshálh* Coast study area, shown in Figure 4.1 in relation to Vancouver, extends from Sechelt in the south to Pender Harbour in the north. To facilitate discussion and analysis, that area was further divided into three study units: Sechelt, Halfmoon Bay, and Pender Harbour. While Sechelt is the region's urban core, Halfmoon Bay and Pender Harbour are popular vacation and retirement destinations.

Both coastal sprawl and climate change are major concerns for the *shíshálh* Nation. The coast is recognized as a site of "unprecedented" population growth:

> The southern portion of our territory [is] experiencing unprecedented population growth, in part as a bedroom community for the rapidly growing urban centre of Vancouver, as a retirement destination and for recreational activity. In 2005, the population of *ch'átlich* (Sechelt) increased by 4.6 percent, almost three times the rate of growth experienced by the Province as a whole. This change has increased the demand for residential

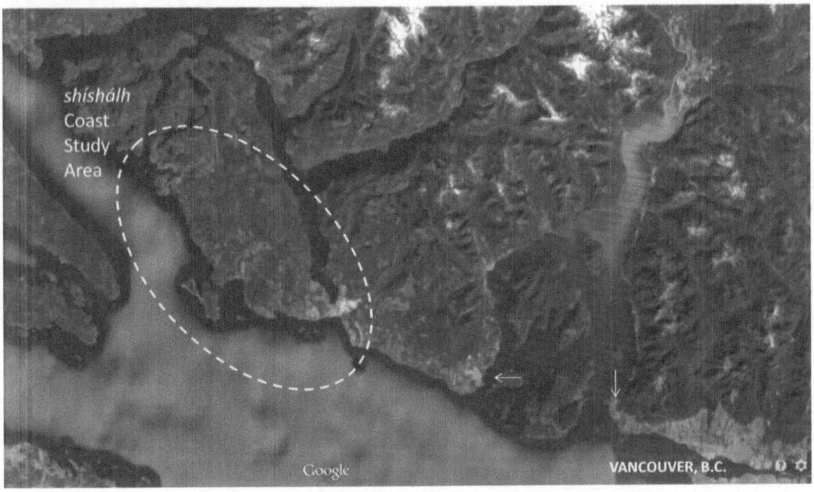

Figure 4.1 The *shíshálh* Coast study area (dashed oval) in relation to the City of Vancouver (bottom-right) and Gibsons (left arrow). Reaching the coast requires a one-hour ferry crossing from Vancouver to Gibsons. The low-lying and thus at-risk community of Squamish is visible at top-right. Vancouver Island is visible at bottom-left, across the Salish Sea. Top is north.

Source: Adapted from Google Maps.

> development, and settlement is now occurring in many areas of *shíshálh* Nation territory at an unprecedented rate. The *shíshálh* Nation is particularly concerned about the potential impacts on cultural sites resulting from this rapid pace of development, and about environmental impacts resulting from the conversion of natural areas to urban areas.
>
> (SIB 2007: 15)

While the Nation recognizes the serious threat climate change poses, it has yet to develop a climate change plan.

In addition to its unique geographic position relative to the regional center of Vancouver, the *shíshálh* Coast study is distinctive in regards to *shíshálh* Nation governance. In 1986, the *shíshálh* entered into a self-government agreement enabling the Nation to obtain control over their lands and resources and enact local by-laws (Etkin 1988; Yellowhorn 1997). Although not explicitly stated, the framework arguably could provide the Nation with greater control over archaeological sites and heritage landscapes (Yellowhorn 1997), recognized by the United Nations (UN) (2008) as integral to the survival of Indigenous peoples. In this light, the *shíshálh* Coast study provides a window into how successful this form of "self-governance" has been as it pertains to the protection of cultural heritage landscapes.

Chapter Four is divided into three parts. I begin by considering the *shíshálh* Nation Strategic Land Use Plan, with specific attention given to cultural heritage and cultural landscapes. In the next section, I examine the problems of coastal sprawl (past and present) and sea level rise (future) as they are unfolding on the *shíshálh* Coast. I end with a summary of key findings.

The *shíshálh* Strategic Land Use Plan

In 2007, the *shíshálh* Nation adopted its Strategic Land Use Plan (SIB 2007). The plan, titled *lil xemit tems swiya nelh mes stutula* (roughly translated as "we are looking after our land, where we come from"), is a living document that sets out the Nation's vision for the long-term future of their territory. Developed with extensive input from its members, objectives of the plan include: articulation of the *shíshálh* vision for the future of their land and their people's place within their territory; clarification of priorities for the use, protection, and management of areas within *shíshálh* territory; and explanation of how the *shíshálh* Nation "assume greater responsibilities for management decisions affecting land and resources" (SIB 2007: 2).

Important background information for the plan is provided in Box 4.1, including an affirmation of *shíshálh* Nation Aboriginal Rights and Title and a profile of the *shíshálh* Nation and Territory (see also Maud 1978; Merchant 2012; Peterson 1990; SIB 2013). For the *shíshálh*, the Land Use Plan represents the Nation's best efforts to summarize their core values and to describe how they would like to see terrestrial and intertidal resources "protected, managed and utilized now and into the future." Those include heritage properties, which include but are not limited to archaeological sites and cultural artifacts including: resting places of *shíshálh* ancestors and remains; sacred and spiritual places; house sites; battle areas; wood and stone fish traps; works of art; travel routes; rock shelters and cave sites; wet sites; and high-altitude sites. A vision statement and guiding principles for *shíshálh* Nation lands and resources are included in Box 4.2.

Box 4.1 Contextualizing the *shíshálh* Nation Strategic Land Use Plan

The *shíshálh* Strategic Land Use Plan begins with this affirmation of *shíshálh* Nation Rights and Title and profile of the *shíshálh* Nation and Territory (SIB 2007: 6–8).

Affirmation of *shíshálh* Nation Rights and Title

The *shíshálh* Nation asserts aboriginal title to and aboriginal rights throughout its territory. We have in the past and continue to engage in cultural activities throughout the territory, activities giving rise to aboriginal rights. We will engage in cultural activities in the territory now and into the future. In addition, we exclusively occupied our territory in and prior to 1846, and assert aboriginal title to all of our territory.

Aboriginal title carries with it the right to choose the use to which the land is put. This Land Use Plan is an exercise of governance by the *shíshálh* Nation over our territory and is intended to provide direction on current and intended future land use and stewardship of the territory.

Profile of the *shíshálh* Nation and Territory

Since time immemorial the *shíshálh* Nation has occupied and utilized its entire territory from the oceanside to the mountain tops utilizing and managing its vast and varied natural resources through the development of complex social institutions, technological innovation and development.

Our territory is located in the southwest corner of what is now referred to as British Columbia, it extends from *xwésám* (Roberts Creek) in the southeast to the height of land located north of *x̱énichen* (head of Queen's Reach) in the north, *ḵwékwenis* (Lang Bay) to the west and *spílḵsen* (Texada Island) to the south.

Prior to European invasion our Nation comprised at least four distinct sub-groups occupying at least twelve large settlements. The groups include:

- *téwánkw* who were located in *stl'íxwim* (Narrows Inlet), *skwúpa* (Salmon Inlet) and *ʔálhtulich* (Sechelt Inlet) and had major town sites located at the head of *stl'íxwim* (Narrows Inlet) and *álhtúlích* (inside waters/Porpoise Bay);
- *x̱eníchen* (Hunechin) with their principal town located at the head of *lékw'émin* (Queen's Reach in Jervis Inlet);
- *ts'únay* (Tsonai) with their main town site located at *ts'únay* (Deserted Bay); and,
- *xíxus* who were located along the outer coast between *xwésám* (Roberts Creek) and *ḵwékwenis* (Lang Bay) with principle town sites located at *ḵálpilín* (Pender Harbour) and *ch'átlich* (Sechelt).

Our Nation practiced a typical North-Coast Salish subsistence pattern utilizing a combination of hunting, fishing and the gathering of shellfish and food plants as they became seasonally available as well as preservation and storage of foods at residential locations. Principal settlements were inhabited by populations who gathered during the winter months and resided in large multi-family dwellings (long houses). The large villages at *ḵálpilín* (Pender Harbour) and *ch'átlich* (Sechelt) were occupied year round by segments of the population. Portions of the population traveled throughout the territory in the warmer months, dispersing and regrouping to make the most efficient use of the available resources. Ungulates and sea mammals formed a significant portion of the diet; however, salmon was the most important food. Much of the food was dried and stored for use during the winter months.

Box 4.2 *shíshálh* heritage principles, part I

The following vision and principles guide the *shíshálh* view of land and heritage, and thus the *shíshálh* Strategic Land Use Plan (SIB 2007: 16).

We, the *shíshálh* people, were put here by the Creator as keepers of our waters and lands. We have lived in our territory according to our own laws and systems of government since time immemorial and will continue to do so forever.

We will give back to the earth the respect and sanctity it rightfully deserves. We will honour our lands, waters and air as our ancestors have taught us. We know how the environment used to be and we will work to rehabilitate our territory's natural resources to what they once were.

The *shíshálh* Land Use Plan represents our vision for the management of our territory—a picture of our desired future—that includes the following:

- Our cultural practices and customs, including those related to the use of land and resources, will be revered and will continue to be handed down from our ancestors through to our Elders and youth with respect. We will preserve *sháshíshálem* (our *shíshálh* language) and continue to promote its use and our cultural ways so that they continue as they have been for countless generations.
- Our leaders will continue to be guided by our cultural values and by the wishes of our community, who will be fully engaged in planning and management for our future.
- Our people will welcome opportunities to accept new ideas and innovations that can assist us in managing our land and resources on a sustainable basis. We will anticipate changes to our territory, and adjust to new challenges such as climate change.
- Our members will be active throughout the territory, and continue to access all areas for economic, subsistence, cultural or other needs. In this way, we will maintain our connection to the land and waters of our territory, and all that it symbolizes for our people. We will resettle some of the village sites that were in place before Europeans arrived.
- Key areas of our territory will be protected from development, to preserve areas of cultural importance so that our land and waters can continue to support healthy populations of wildlife and we ourselves as a people that depend on them for our way of life. The natural ecosystem processes that have occurred over thousands of years in our territory will continue unimpeded.
- We will adopt ecosystem based management approaches that reflect our understanding of the connection among all things, and the need for planning over extended timeframes and at multiple scales.

- Our decisions regarding land and resources will continue to reflect our humility and connection to all things, and our commitment to sustainability for current and future generations. Our decisions and actions will make clear that our interest in resource development is not driven simply by economics, but by sustainability for all people that choose to make our territory their home.
- We will secure greater authority over the management of land, water and resources within our territory. Land within our territory will no longer be alienated from us, and we will secure compensation for resources that have been removed without our consent. Overharvesting and destructive resource extraction will cease.
- We will achieve greater self-sufficiency as a people, and will sustain ourselves with more of our own materials and energy so that we can thrive within our territory and reduce our dependency on others for our well-being.
- Our communities will be safe, secure and healthy. Our young people will achieve health and educational standards that compare favourably with the rest of Canada.
- We will establish cooperative relationships with those who have chosen to make our territory their home, and with commercial and industrial interests that operate within our borders.
- Sustainable industry and commerce will thrive in our territory, and operate in a manner that assures the long term health of the land, resources and our people as well as the economic well-being of our people.

Over the years, the *shíshálh* have seen their land and heritage developed without their consent. This is described in the Land Use Plan (SIB 2007: 1), which notes that "[a]reas that have great cultural and historical importance for our people have been impacted, and access to some areas that we have used for generations has been denied. Particularly in recent years, development pressures within our territory have increased," including forestry, residential and foreshore development, and industrial development.

As development pressures have increased, the *shíshálh* Nation has been inundated with requests from Government agencies to assess proposals for projects through the referral process, which often requires evaluating heritage sites. Unfortunately, "more often than not, the referral is received because the site is in danger of being altered, damaged or destroyed by development" (Union of British Columbia Indian Chiefs [UBCIC] 2013: 9). As described in the Land Use Plan (SIB 2007: 1), this approach puts the *shíshálh* Nation in a "reactive position," placing a "significant burden" on *shíshálh* technical staff as they attempt to "respond meaningfully to every project on an individual basis." As made clear in the Plan:

> The referral process also fails to address fundamental questions regarding our rights and title and leaves our Nation without a meaningful role in planning or management decision-making. Through such experiences, we have come to understand the risks of coping with development pressures one project at a time. We have developed this Land Use Plan in order to provide a more comprehensive and integrated view of our territory, so that we can be proactive in determining what happens in the future.
>
> (SIB 2007: 1–2)

The *shíshálh* Nation has concerns over both the "rapidly growing population" as well as the "adequacy of current housing available, particularly for the elders and for those with lower incomes" (SIB 2007: 86). Their concern about unsustainable population growth is borne out in Figure 4.2, which shows historic and expected population growth for the period 1956–2036. The population of the region sat around 5,000 people in 1950. It will be approaching 50,000 in 2050, a growth rate of 1,000 percent per century, or 10 percent per year.

The protection of cultural heritage is a priority for the *shíshálh* Nation. Generally speaking, *shíshálh* approaches to heritage and heritage stewardship are holistic, ecosystem-based, human-centred, and built on community and learning; that is, it is "adaptive" (Armitage et al. 2007; see also Berkes 2008; Menzies 2006; Rose 2005). That approach is illustrated best in the Nation's guiding principles for planning and management of land and resources, included here as Box 4.3.

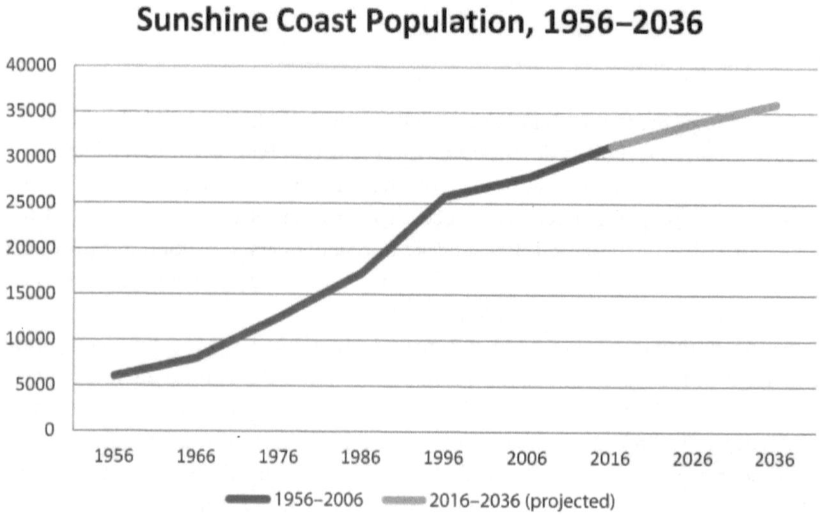

Figure 4.2 Sunshine Coast population, 1956–2036.

Source: Data from BC Stats (2011a).

Discussion of cultural heritage in the Land Use Plan is prefaced by the following statement:

> Since time immemorial the *shíshálh* Nation has occupied our territory. A consequence of this occupation are the cultural properties (e.g., artifacts, spiritual sites, stories, names and traditions) left behind by our ancestors. It is a central principle of the *shíshálh* Nation Heritage Policy that all cultural properties belong to those who made them regardless of the world—living or spiritual—within which they live. Therefore, the *shíshálh* Nation has primary jurisdiction to manage, protect and preserve the history of the Nation throughout our entire territory. The management of *shíshálh* cultural resources must reflect *shíshálh* values for the purpose of preserving and protecting *shíshálh* culture, and to ensure that our traditions and way of life live on.
>
> For over two hundred years the needs of settlers and development have superseded the protection of *shíshálh* cultural resources, resulting in the destruction of our history. Such development continues today and threatens the existence of our culture.
>
> The necessity to protect our culture should be primary over the needs of external development. Therefore all land use within our territory must be planned so as to result in as little conflict as possible with *shíshálh* cultural resources.
>
> (SIB 2007: 18)

Box 4.3 *shíshálh* heritage principles, part II

The following guiding principles for planning and management are laid out in the *shíshálh* Strategic Land Use Plan (SIB 2007: 17):

- *Guidance of Ancestors and Elders*: All *shíshálh* Nation land use planning and resource management activities shall be undertaken in a manner that is consistent with the teachings taught to us by our Ancestors and Elders.
- *Respect*: *shíshálh* Nation use and management of the land shall always reflect our deepest respect for the land and the interconnectedness of all things.
- *Sustainability*: *shíshálh* Nation use and management of resources shall be guided by our commitment to sustainability both in the present and for all generations to come—which means maintaining diverse and abundant wildlife and ecosystems in perpetuity while providing for diverse cultural, social and economic activities that support a balanced, healthy, secure and sustainable quality of life.

- *Self-Reliance*: Our use and management of land and resources shall seek to increase the self-reliance of the *shíshálh* Nation, so that we can support our own communities and others that have chosen to make our territory their home over the long term.
- *Cultural Practices*: *shíshálh* Nation planning, use and management activities shall ensure that access is maintained for our people so that our cultural practices can continue unimpeded throughout our territory.
- *Cultural Sites*: Sites of our past, current and future use and occupation of the territory shall be respected and preserved.
- *Ecosystem-Based Approach*: Management of our *shíshálh* Nation territory shall adopt a holistic, ecosystem-based approach that considers the entire ecosystem in determining use of specific areas and setting harvest rates that focus on what to leave behind, rather than what to take.
- *Capacity Building*: We shall strive to provide opportunities for members of the *shíshálh* Nation, particularly the youth, to build their skills and experience in management of lands and resources, so that they and all future generations continue to act as stewards of our territory.
- *Knowledge*: *shíshálh* Nation planning and management approaches for land and resources shall incorporate both our own cultural and local knowledge as well as western science-based understandings.
- *Space and Time*: *shíshálh* Nation planning and management approaches shall take into account multiple spatial scales and time frames, and seek to maintain or increase resilience in the face of critical and long-term issues such as climate change.
- *Precaution*: A precautionary approach shall be adopted for land planning and management, so that decisions err on the side of caution when information is limited.
- *Consent*: Development of land and resources shall only proceed when the risks of impacts on our territory are well understood and accepted by the *shíshálh* Nation.
- *Benefits*: The *shíshálh* Nation shall benefit fairly from development and use of land and resources within our territory.
- *Monitoring*: The condition of the land and resources across *shíshálh* Nation territory shall be monitored, and knowledge of trends and responses to change shall be incorporated into future decision-making through adaptive management.

It is clear from the Strategic Land Use Plan that significant changes are taking place on the *shíshálh* Coast, including but not limited to intensive residential and marine development. That these changes constitute a major concern for *shíshálh* people is also clear. While climate change is recognized as a serious threat, it has not yet been integrated into *shíshálh* land use strategies; this includes sea level rise. Two key points need highlighting before proceeding to the survey. The first concerns *shíshálh* Nation's authority over the stewardship of their heritage, and the second, the relationship between cultural survival and external development. As shown in Table 4.1, the Strategic Land Use Plan lays out a goal and three "priority actions" concerning *shíshálh* Nation control over the stewardship process.

A useful frame for approaching Table 4.1 is the poignant concept "developed without consent." In such light, the goal of ensuring *shíshálh* authority over planning and management can be seen as a response to the historic loss (or taking) of control at the hands of outsiders (or colonialists), a wholly asymmetrical power relationship that clearly continues into present times. One manifestation of that relationship on the landscape is development, specifically "external development."

In the context of the *shíshálh* Coast, *external development* refers to development authorized and undertaken by outsiders and settlers. External development has become such a problem that some *shíshálh* members have expressed concern over the "loss of future options for the *shíshálh* Nation given the pace of private development in prime settlement areas, particularly along the waterfront" (SIB 2007: 86). As illustrated in the next section, these concerns are justified, indeed even understated in light of sea level rise.

The *shíshálh* Coast Study

The *shíshálh* Coast study is an exploration of how the maritime heritage crisis is playing out on British Columbia's famous "Sunshine Coast." My study was guided by the themes identified in Chapters Two and Three of this book. Those themes, expressed here as questions, are:

1. What is coastal change?
2. What are the impacts of such changes on heritage?
3. What has been the response to such change?
4. How effective is this response?

Table 4.1 *shíshálh* Nation goals and priority actions concerning authority over the heritage stewardship process

Goals	*Priority actions*
Ensure that the *shíshálh* Nation has authority over the planning and management of cultural resources in the territory.	• Develop and implement *shíshálh* laws, regulations, guidelines, and/or protocols to protect *shíshálh* archaeological and cultural sites, cultural practices, and intellectual property. • Require archaeological impact assessments prior to development in areas of known or probable archaeological potential. • Require *shíshálh* approval of all site alteration permits under the British Columbia *Heritage Conservation Act.*

Source: SIB 2007: 19, Table 2.

To address these questions in the context of *shíshálh* First Nation heritage, my study was informed by the Rapid Coastal Zone Assessment Survey (RCZAS) (see Chapter Three), which provides a framework to evaluate the past, present, and future impacts of coastal change on archaeological sites. I evaluated (*i*) the number of archaeological sites, (*ii*) the number of sites located at less than 15-meter elevation, (*iii*) the number of sites that intersect the shoreline, (*iv*) the number of sites "disturbed" and "destroyed," and (*v*) the kinds of disturbances operating in each area. These data were compiled using archaeological site information available through the British Columbia Archaeology Branch. This aspect of my study constitutes a portion of the RCZAS desktop survey (Phase One).

I limited the field portion of the study to a landscape survey, documenting examples of coastal change in the three sub-areas that are Sechelt, Halfmoon Bay, and Pender Harbour, shown in Figure 4.3. Together, these approaches enabled me to assess the success of CRM in mitigating coastal change impacts, and to consider their social consequences.

The *shíshálh* Coast study was carried out not just with the Strategic Land Use Plan in mind, but also with consideration for previous research into the effects of development on *shíshálh* heritage. In 1976 and 1977, archaeologists Steve Acheson and Syd Riley conducted regional shoreline surveys and inventories in three large sections of the northeastern Salish Sea (Acheson and Riley 1979; see also Welch et al. 2011: 95–8). The primary objective of their inventory was to establish a database to "facilitate future planning and management of the

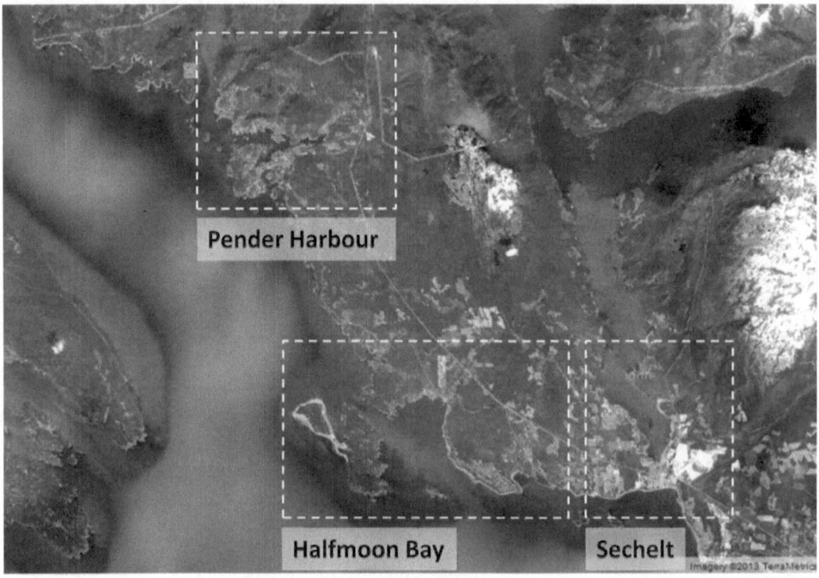

Figure 4.3 To facilitate analysis, the *shíshálh* Coast study area was divided into three discrete Study Units: Sechelt, Halfmoon Bay, and Pender Harbour. Top is north.

Source: Adapted from Google Maps.

heritage resource within the Gulf of Georgia in response to intensified land-use and demographic pressures taking place" (1979: 39). In all, Acheson and Riley compiled information on 186 sites located along 156 kilometers of shorelines. Most importantly, they quantified modern (pre-1977) landscape disturbances and made predictions about future impacts (post-1977). One of the areas visited by Acheson and Riley, Smuggler Cove, falls within this study's Pender Harbour Study Unit, as discussed below.

Three decades after Acheson and Riley's study, archaeologist Peter Merchant (2004) undertook an intensive survey of 16 kilometers of the Pender Harbour coastline, asking similar but more detailed questions about the nature of heritage destruction here. In all, Merchant compiled information on and made recommendations for 19 sites.

As discussed below, these two studies established important historical baselines when it came to quantifying changes in coastal change impacts over time. I begin by describing the southern-most *shíshálh* Coast Study Unit, the area surrounding the City of Sechelt. I then work my way "up-coast" to Halfmoon Bay, ending in the north at Pender Harbour. For each study unit, I provide photographs and descriptions of the local landscape. I then quantify the results of my survey.

Sechelt Study Unit

The Sechelt Study Unit is the most heavily developed area considered in this investigation (Figure 4.4). Its heavy footprint reflects the area's role as the commercial center for the central Sunshine Coast. In addition to being a transportation center, the area is also the nexus of political power, as it is the headquarters for three regionally significant governing bodies: the District Municipality of Sechelt, the Sunshine Coast Regional District, and the *shíshálh* First Nation. While the District of Sechelt is situated on top of the *shíshálh* Nation's principle town site of *ch'átlich*, the Sunshine Coast Regional District covers nearly perfectly the territory of the *shíshálh* Nation.

Development is central to the economy of the District of Sechelt, which is largely dependent on the summer tourism industry, a booming retirement population, and the area's growing popularity as a commuter town. This follows a long history of industrial resource extraction, notably logging (Peterson 1962).

Much of the low-lying southern shoreline in Sechelt is protected by a seawall, yet there is a strong desire to develop these already vulnerable coastal areas (Figures 4.5 and 4.6), irrespective of potential coastal erosion threats, real or perceived. There is little, if any, unmodified shoreline to be found in the Sechelt area.

In addition to seawalls, land uses in this area include marinas (boat and seaplane), residential housing, commercial (e.g. restaurants), industry (e.g. gravel quarry), and recreational (e.g. parks). In the Sechelt Inlet area just north of the town itself, development continues to encroach upon forests (Figure 4.7), and coastal erosion of heritage sites is prolific (Figure 4.8). "Natural" heritage is privileged (Figure 4.9) over cultural heritage, and residential development is slowly creeping up the inlet, resulting in substantial landscape modification (Figure 4.10).

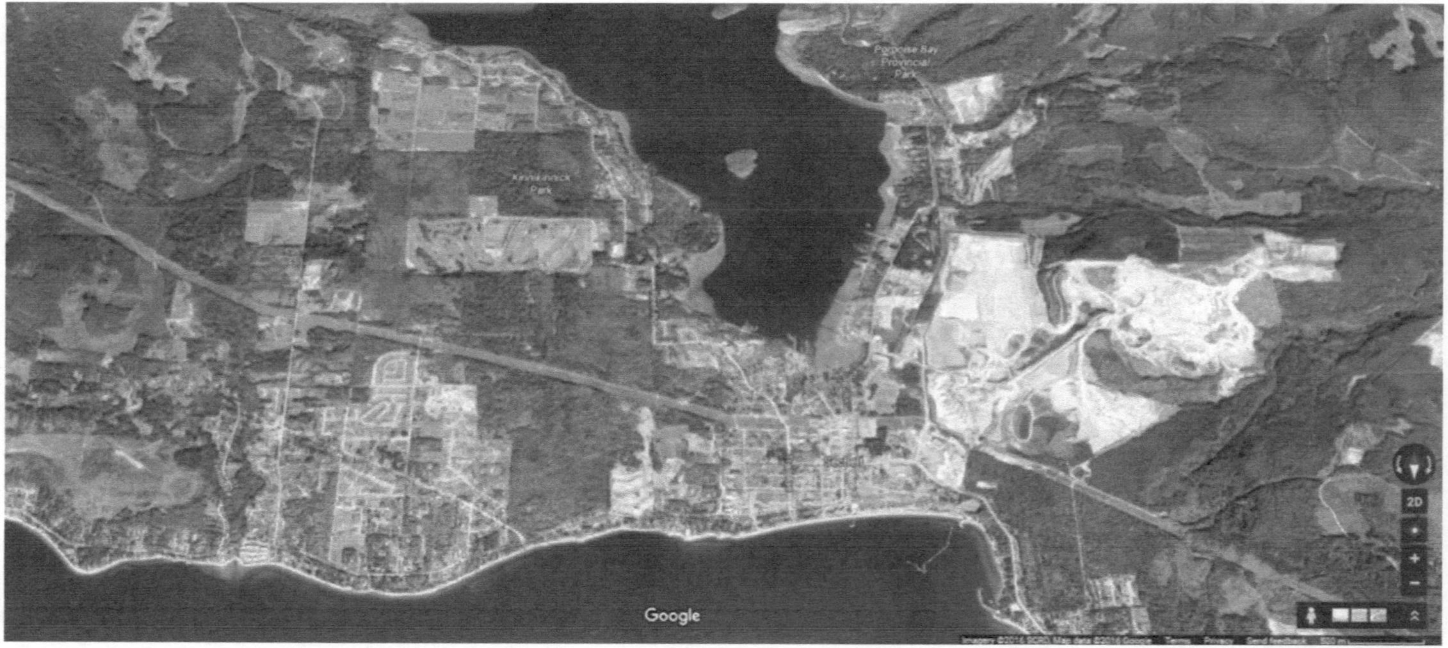

Figure 4.4 Satellite image of the heavily developed Sechelt Study Unit. The water body to the south is the Strait of Georgia; Sechelt Inlet, which includes the Angus Creek and Snake Bay study areas (top-right and top-left, respectively) is to the north, located behind the low-lying isthmus which defines Sechelt's downtown core. Top is north.

Source: Google Maps.

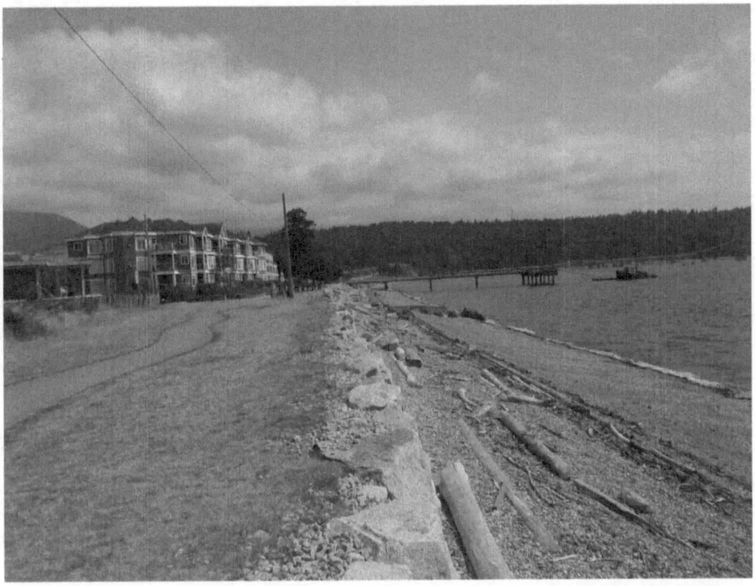

Figure 4.5 Trail Bay seawall and promenade near Marine Gateway Park and Sechelt Wharf in the Sechelt Study Unit. The building on the left is a condominium. This image illustrates the degree to which the shoreline has been developed and its exposure to the Strait of Georgia.

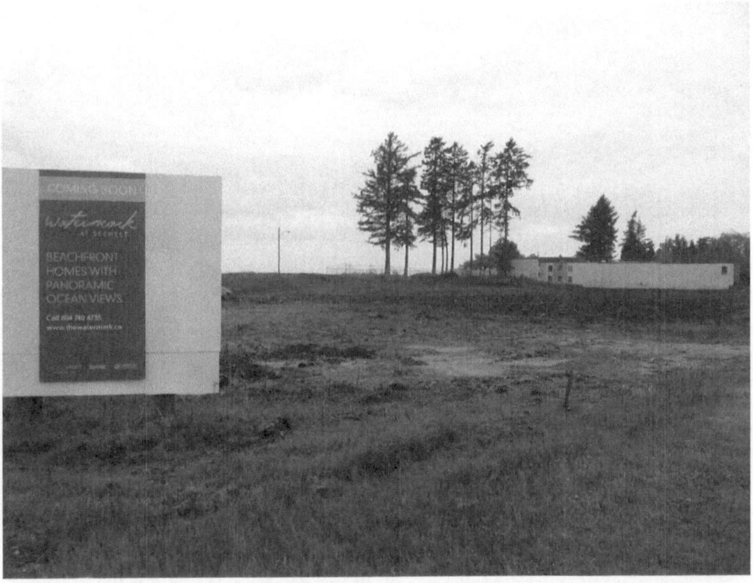

Figure 4.6 Heritage protection at the Watermark condominium project included the fencing-off of this lone stand of trees in the southwest corner of the lot, where the already mostly destroyed archaeological site includes/included wet site, lithic, and shell midden components. The Trail Bay seawall is behind the berm at centre between the sign and the trees.

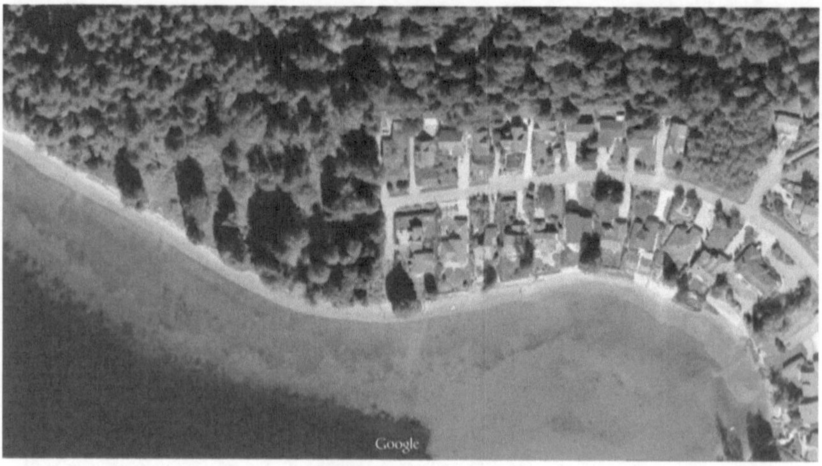

Figure 4.7 Satellite image of eastern Sechelt Inlet illustrating the different "management" practices, where *shíshálh* First Nation land is to the left and a developer's vision on the right. Top is north.

Source: Google Maps.

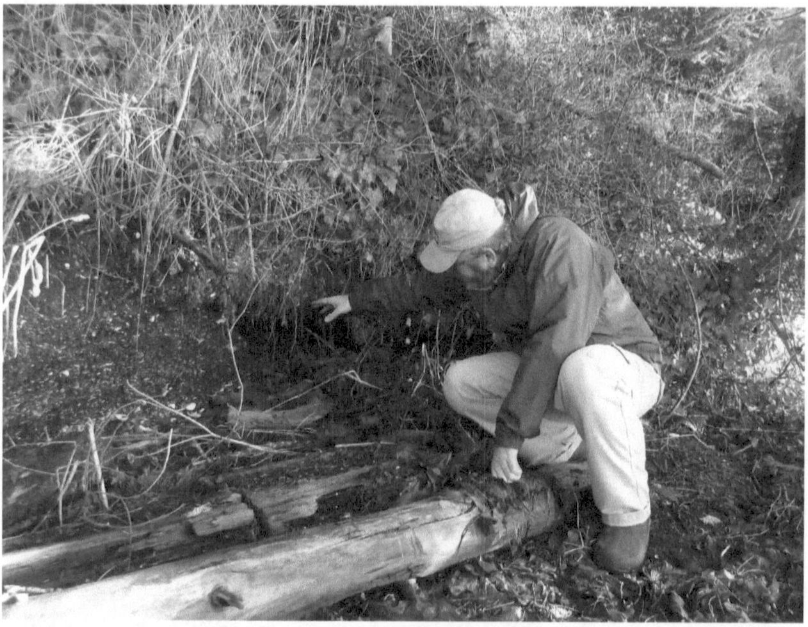

Figure 4.8 Wave erosion undercutting site in eastern Sechelt Inlet. While logs may protect shell midden sites in some contexts, when afloat at high tide they act as giant abraders and pile drivers, scouring the shell midden and destabilizing the bank, as shown here. This entire landscape (see Figure 4.7) is at significant risk due to development and sea level rise.

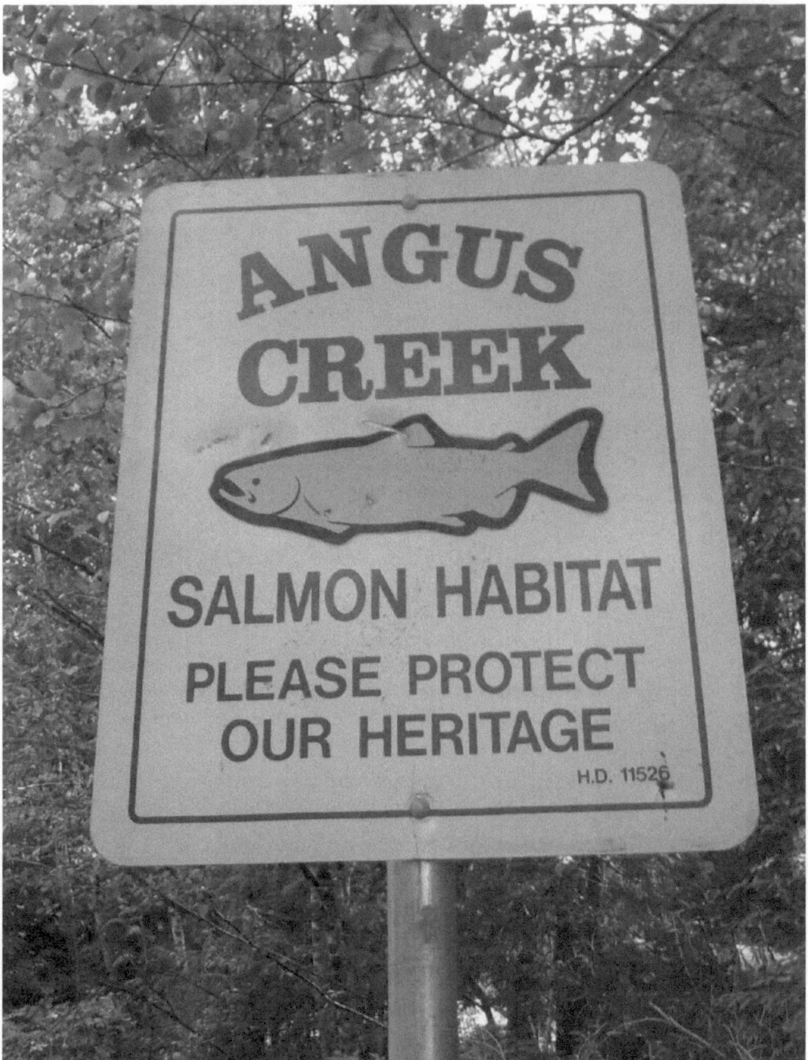

Figure 4.9 Heritage signage in eastern Sechelt Inlet. Sign reads "ANGUS CREEK SALMON HABITAT—PLEASE PROTECT OUR HERITAGE." This is symbolic of how "heritage" is both natural and cultural, but selectively represented and thus differently prioritized. This roadway sign is within a major *shíshálh* heritage landscape.

Halfmoon Bay Study Unit

The Halfmoon Bay Study Unit is located northwest or "up-coast" from the Sechelt Study Unit (Figure 4.11). Interestingly, the colloquial up-coast/down-coast distinction defines one's geographic location on the Sunshine Coast in

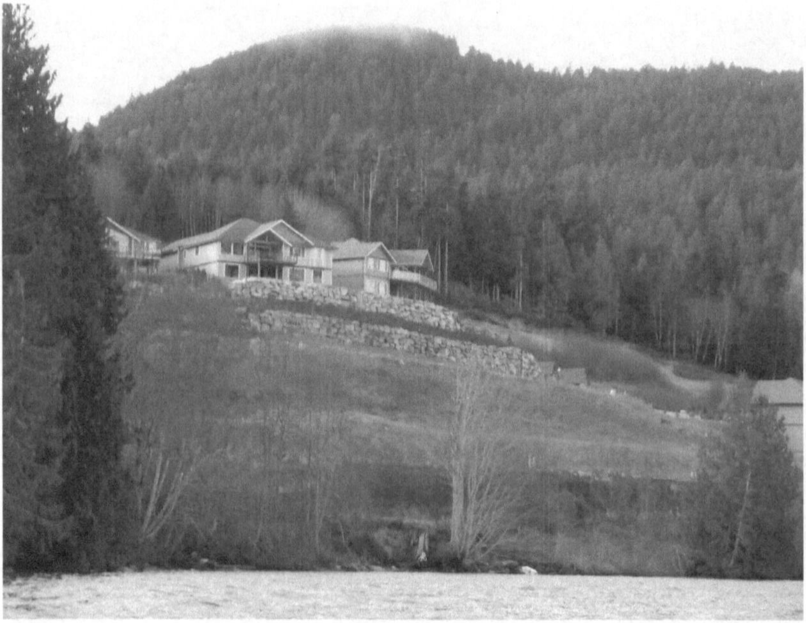

Figure 4.10 Close-up of intensive landscape modification in west Sechelt Inlet. This image illustrates legal and permitted coastal development immediately adjacent to a marine delta and *shíshálh* heritage landscape. View is southwest.

relation to the City of Vancouver, specifically the ferry terminal at Horseshoe Bay. The further one is "up" the coast, the further "away" from Vancouver one is.

A 15-minute drive from downtown Sechelt to southern Halfmoon Bay along the Sunshine Coast Highway, there exists a very clear urban–rural distinction. During my year-and-a-half residence in Halfmoon Bay for this study, downtown Sechelt came to be associated with driving and commercialism, and the intervening space simply became a blurred travel corridor. In opposition to those feelings, my return trip to Halfmoon Bay became associated with rural retreat and escape. This dynamic was far stronger in relation to required trips into Vancouver, colloquially known as the "big city." That far more intensive trip, which minimally required at least a three-quarter-day investment, involved car, ferry, and bus. Along with its many "million dollar" views, the rural maritime character of the Halfmoon Bay Study Unit distinguishes it most from down-coast Sechelt, and Vancouver beyond that.

Halfmoon Bay is known for its exclusive ocean view properties (Figure 4.12). Despite this, the number of public beach access points is limited. This is due, in part, to the steep bluffs that predominate in the south, and to the intensive and extensive nature of private shoreline development. The Halfmoon Bay Study Unit features Smuggler Cove Provincial Marine Park, a popular destination for boaters heading up-coast to Pender Harbour (Figure 4.13).

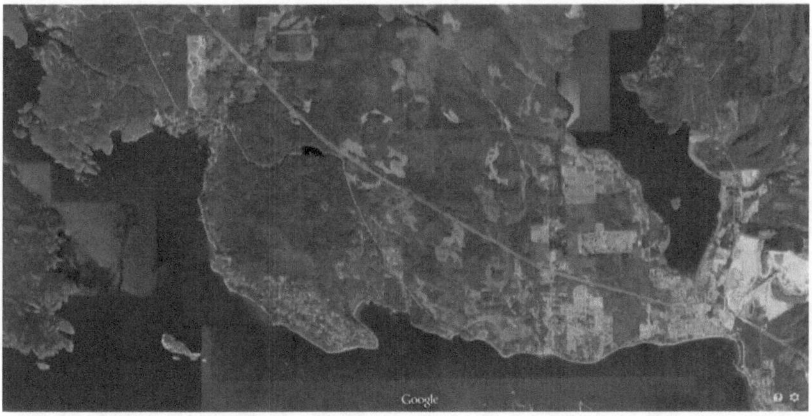

Figure 4.11 Idyllic Halfmoon Bay Study Unit (left) in relation to the much more urban Sechelt Study Unit (right). Areas of light coloration indicate more intensive development. Top is north.

Source: Google Maps.

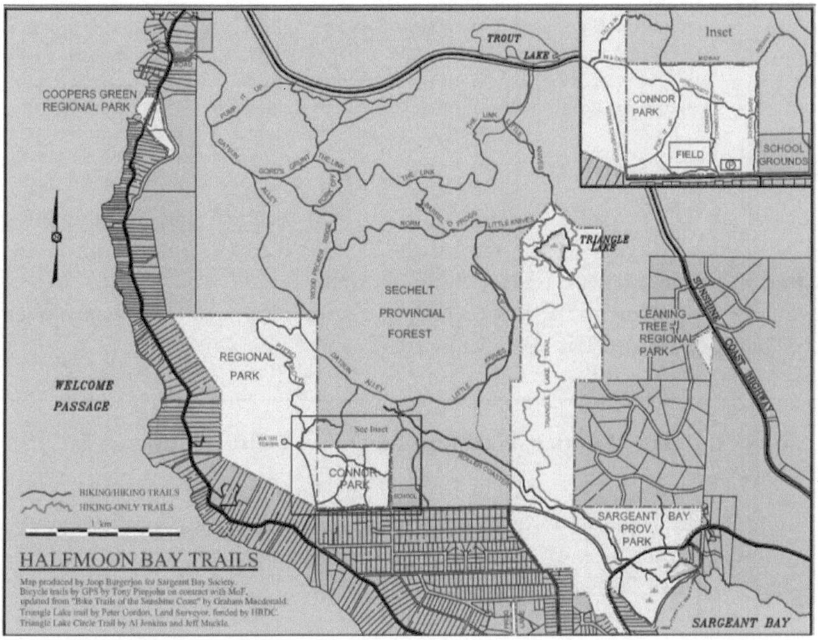

Figure 4.12 Halfmoon Bay recreation map. While the hilly and forested interior landscape is protected by a number of provincial, regional, and local parklands, the coastal strip along scenic and highly valued Redrooffs Road has been heavily developed, as indicated by the numerous property divisions.

Source: Sargeant Bay Society, reprinted with permission.

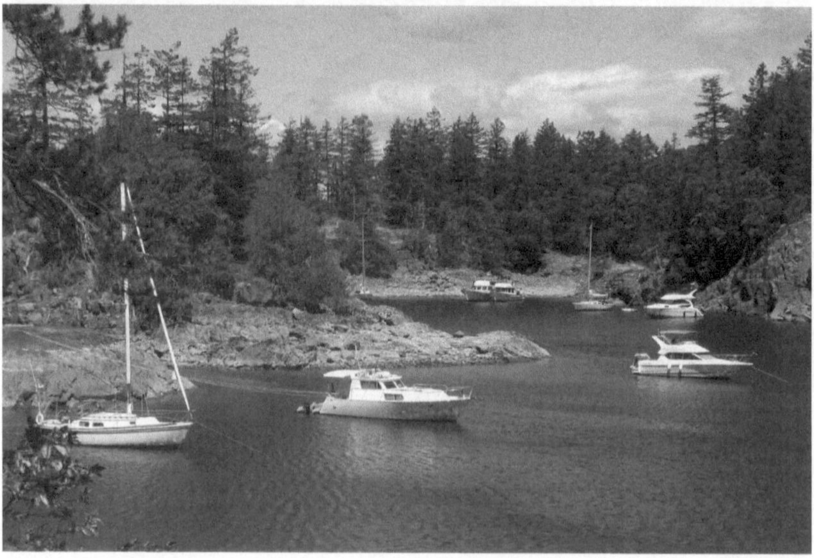

Figure 4.13 Yachts moored in Smuggler Cove Marine Provincial Park, northern Halfmoon Bay Study Unit. Although Smuggler Cove is located a distance off the main highway and requires a half-hour walk to reach the sea, boats allow easy access. Recreational boating represents a complicating factor when it comes to issues of coastal resource management (see Figure 4.14).

Pender Harbour Study Unit

Located about 35 road kilometers up-coast from Sechelt, and 60 kilometers (120 minutes driving) from Gibsons, the Pender Harbour Study Unit is the ultimate seaside getaway (Figure 4.14). For the *shíshálh*, this cultural area is extremely important and at-risk (Box 4.4), a point illustrated by concerns over *séxwʔámin*, or Garden Bay (Box 4.5).

Box 4.4 The Pender Harbour–Sakinaw Cultural Emphasis Area

The Pender Harbour Study Area is part of the Pender Harbour–Sakinaw Cultural Emphasis Area, described in the *shíshálh* Nation's Strategic Land Use Plan (SIB 2007: 77–8) as follows:

Area Description

The *kálpilín – stséxwena kw'enit sim alap* (Pender Harbour – Sakinaw Cultural Emphasis Area) comprises approximately 10,623 hectares and contains an extremely high concentration of *shíshálh* cultural use and occupation sites, cultural features and archaeological evidence. Located

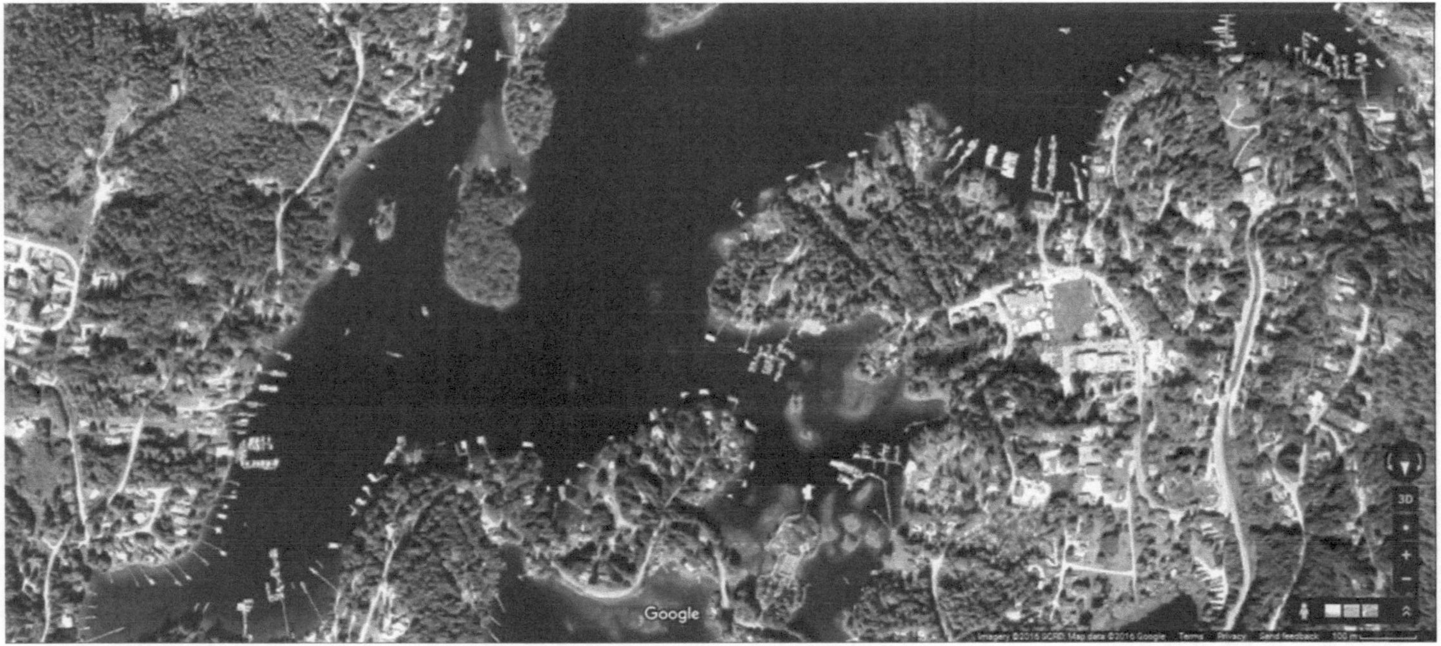

Figure 4.14 Detailed satellite view of coastal residences and docks in the central Pender Harbour Study Unit. Top is north.

Source: Google Maps.

at the heart of the historically most populous region in the territory, this area was the site of the main winter villages of the *shíshálh* people, and included a great many well-protected home sites and productive harvest locations with varied marine and terrestrial resource opportunities. To the north, an extensive lake district was used for hunting and fishing with main camps at *stséxwena* (Sakinaw Creek) and *kwíkwilúsin* (east side of Sakinaw Lake) and *lóh-uhlth* (Mixal Lake).

Key Values, including:

- Extremely high cultural and spiritual values, including:
    - Very high concentration of *shíshálh* cultural use, occupation, and archaeological sites
    - *séxw?ámin* (Garden Bay) could be considered the "centre of the *shíshálh* universe"; it was a year round village by virtue of its sheltered location, and served as a gateway to transportation corridors on *sinkwu* (Georgia and Malispina Strait) and *lékw'émin* (Jervis Inlet) via *líkw'émin* (Agamemnon Channel);
    - Includes main winter villages of *poke-poke-um* (Bargain Harbour), *sálálus* (Madeira Park), *smexhalin* (Kleindale), *kway-ah-kuhl-ohss* (Myer's Creek), and *séxw?áwinl* Garden Bay);
    - Primary location for winter dances and ceremonials;
    - Many lakes on upper Sechelt Peninsula were favoured harvest sites;
    - Fishing at Ruby, Ambrose and *stséxwena* (Sakinaw) lakes for Rainbow Trout, and hunting, primarily for *húpit* (Black Tailed Deer), in adjacent forests;
    - Wide variety of seasonal settlements and camps to facilitate harvesting of local resources.
    - *wah-wey-we'-lath* (Mt Cecil) and *shélkém* (Mt Daniel) were important local mountain peaks used for a variety of cultural and spiritual purposes and in defense;
    - Numerous fish weirs, canoe skids, ceremonial sites, archaeological sites.
- High wildlife/biodiversity values, including:
    - Extensive and intensive foreshore, inter-tidal and marine harvesting of resources.

Key Management Issues, including:

- Proliferation of docks, boats and foreshore development.
- Land alienation due to intensive rural residential development.
- Water pollution from septic systems.
- Shellfish contamination.

- Loss of access to cultural harvesting sites and resources.
- Potential impacts of water diversion from Sakinaw/Ruby lakes for domestic residential use.
- Destruction of *shíshálh* cultural resources and archaeological sites.

Management Direction, including:

- To be developed through a Cultural Emphasis Area management planning process.
- Detailed landscape level planning and zoning required to protect and restore opportunities for *shíshálh* harvesting, especially in areas of current or proposed development expansion.

Box 4.5 *séxwʔámin* (Garden Bay) and the Pender Harbour study area

Central to the Pender Harbour–Sakinaw Cultural Emphasis Area is *séxwʔámin* (Garden Bay), described in the *shíshálh* Nation's Strategic Land Use Plan (SIB 2007: 9–10) as follows:

In social, political and economic terms *sé<u>x</u>wʔámin* (Garden Bay) could be considered one of the most important residential locations of the *shíshálh* Nation and is an example of the Nation's intense and complex utilization of its territory. Located at the heart of one the most populous regions in the territory, it dominated *<u>k</u>álpilín* (Pender Harbour) and the surrounding area. Together with the villages of *p'úkwp'a<u>k</u>wem* (Bargain Harbour), *sálálus* (Madeira Park), *smishalin* (Kleindale) and *kway-ahkuhl-ohss* (Myer's Creek), it shared well-protected home sites and productive harvest locations with varied marine and terrestrial resource opportunities.

> The winter homes of the Sechelt (sicelt) were in Pender Harbour. Near the hospital on the present-day reserve was *tsxwaman*. There were seven big houses there; four of them stood one behind the other while the other three ran crosswise close to the ends. Farther away, on each side of the harbour, were the small houses.

Situated just south of an extensive lake district, it [*séxwʔámin*] was further supported by strategic camps at *stséxwena* (Sakinaw Creek) and *kwíkwilúsin* (east side Sakinaw Lake). *Séxwʔámin* was a year round village that by virtue of its location, served as the gateway to transportation corridors on *sínkwu* (Georgia Strait and Malaspina Strait) and *lékw'émin* (Jervis Inlet) via *lílkw'émin* (Agamemnon Channel). Together with the village of *klay-ah-kwohss* (Buccaneer Bay) on *sxwélap* (Thormanby Island) to the south, it also served as a primary location for winter dances and ceremonials.

> ... at SAUGH-KWAH'-MAIN, [*séxw?ámin*] Garden Bay. Each chief village, he said, had its lodges there, reserved for winter ceremonials.
>
> The Johnson Reserve at Garden Bay, likewise, in the thick layers of shell exposed along the bank, shows signs of long occupancy. Present day residents cannot recall the old buildings, but, from the position of the clam heaps, and from the contours of the terrain, it would seem logical to suppose that they were located on the flat ground immediately above the steep bank to the salt water, at an elevation of some forty feet. The stream that drains what is now Garden Bay Lake runs through that flat.

With the increase in European settlement in *kálpilín* (Pender Harbour) during the early 20th century, many resident *shíshálh* Nation members were forced to move to *séxálíten*, the harbour entry islands (Skardon Islands). Apart from small remnant populations at some of the old village sites, little evidence remained of the once vibrant community that prospered there.

As detailed by Peter Merchant (2004), development and tourism-related impacts here have been a great concern for the *shíshálh*. Merchant's survey of 16 kilometers of southern Pender Harbour shoreline identified 18 archaeological sites. Of the 18 sites, 16 are above 60% destroyed (2004: 53–5). These landscapes offer excellent examples of the kinds of development pressures that typify this area, from sea to sky (Figure 4.15). In addition to terrestrial disturbances associated with residential development and infrastructure, "marine disturbances" (i.e. docks) are identified as a major problem (Figure 4.16). I return to Merchant's study later in this chapter.

Results

In 2014, there were 129 registered archaeological sites identified within the *shíshálh* Coast study area. With 71 sites, the Pender Harbour Study Unit has more than the Sechelt and Halfmoon Bay Study Units combined (Table 4.2).

As predicted, archaeological sites in the urban Sechelt Study Unit have been impacted noticeably more than in outlying areas. A total of 84 percent of known sites in the Sechelt Study Unit are characterized as disturbed or destroyed. This is, on average, ten percentage points higher than at Halfmoon Bay (73 percent)

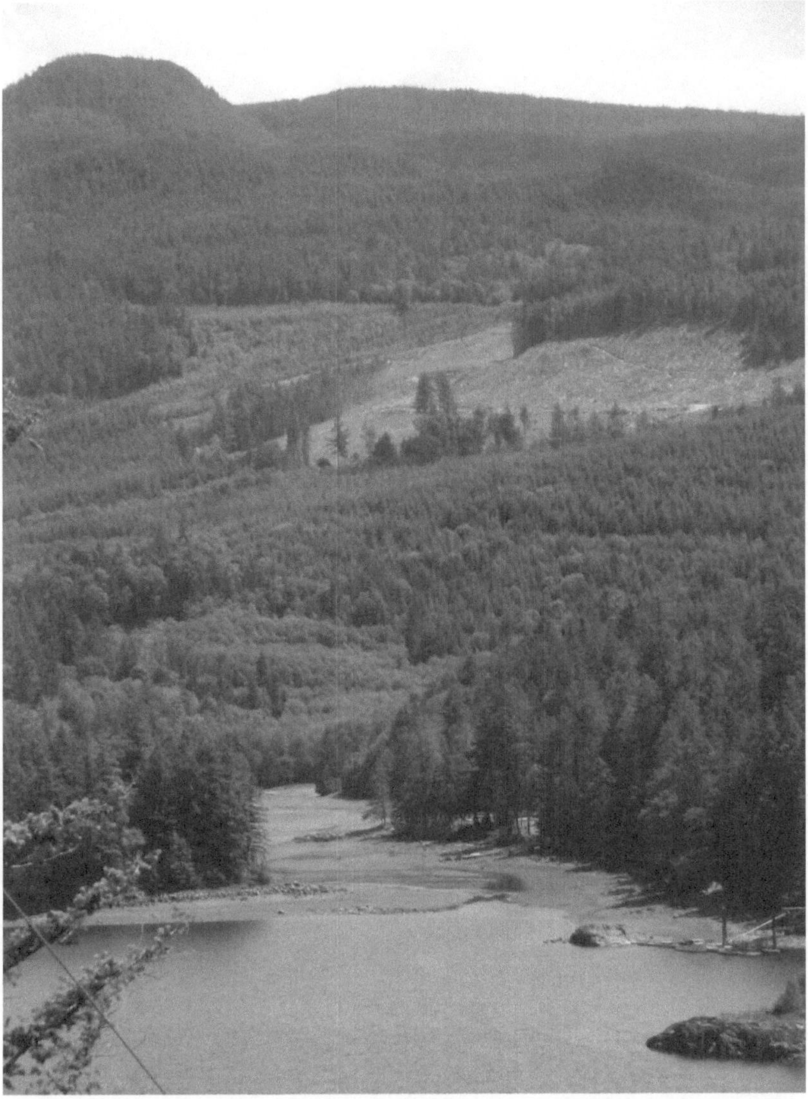

Figure 4.15 Eastern boundary of Pender Harbour Study Unit. While landscapes in the low-lying eastern bays are at risk from development and sea level rise, activities in the upper watershed are also a factor, in this case clear-cutting of the forest.

and Pender Harbour (75 percent). A major factor in this difference is the high percentage of destroyed sites in the Sechelt Study Unit (Figure 4.17).

Three-quarters of all sites in the *shíshálh* Coast study area are described as being disturbed or destroyed (Figure 4.18); about 71 percent are disturbed,

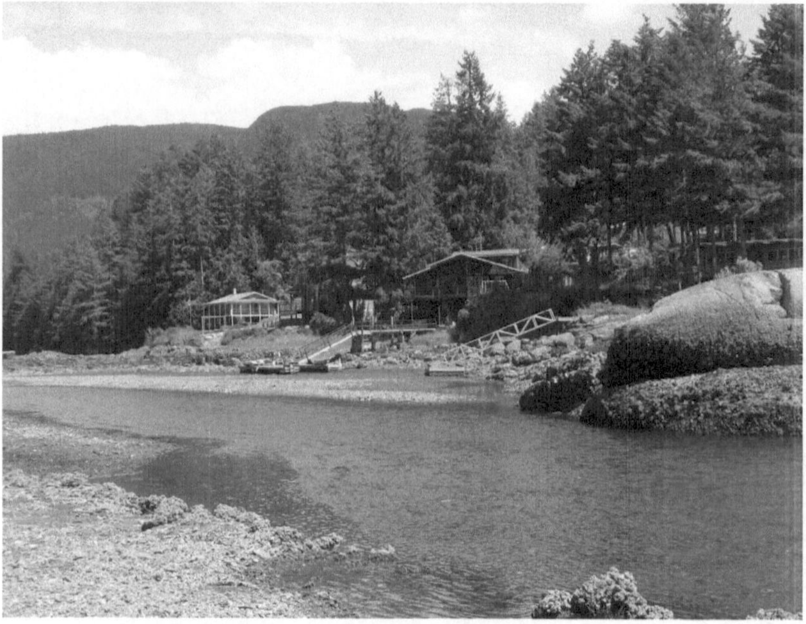

Figure 4.16 Residential development in eastern bay of Pender Harbour Study Unit. Boat ramps and docks act as gateways to the entire marine landscape.

Table 4.2 Summary comparison of three *shíshálh* Coast Study Units

	SSU		*HBSU*		*PHSU*		*Total #*	*Total %*
	#	%	#	%	#	%		
Number of known sites	25		33		71		129	100
Disturbed	19	76	24	73	48	68	91	71
Destroyed	2	8	0	0	5	7	7	5
Disturbed and destroyed	21	84	24	73	53	75	98	76
< 15-meters elevation	16	64	32	97	70	99	118	91
Intersects shoreline	15	60	28	85	70	99	113	88

SSU: Sechelt Study Unit; HBSU: Halfmoon Bay Study Unit; PHSU: Pender Harbour Study Unit.

5 percent destroyed, and 24 percent categorized as "unknown." This number is considered a minimum estimate because of (*i*) the high variability in details included in archaeological site forms and (*ii*) the ontological problem of "site disturbance" versus "site destruction" (see below).

Residential development is by far the biggest threat to *shíshálh* heritage today (Figure 4.19); indeed, residential development has impacted more than twice the number of sites than any other factor. This finding lies in contrast to Acheson and Riley's 1977 identification of logging as the most common cause

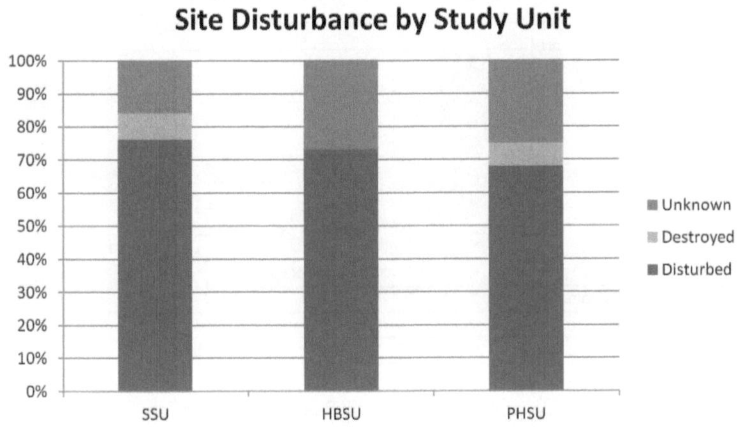

Figure 4.17 Percentage of disturbed and destroyed sites by *shíshálh* Coast Study Unit (SSU: Sechelt Study Unit; HBSU: Halfmoon Bay Study Unit; PHSU: Pender Harbour Study Unit) (n = 129).

Source: Data from British Columbia Archaeology Branch.

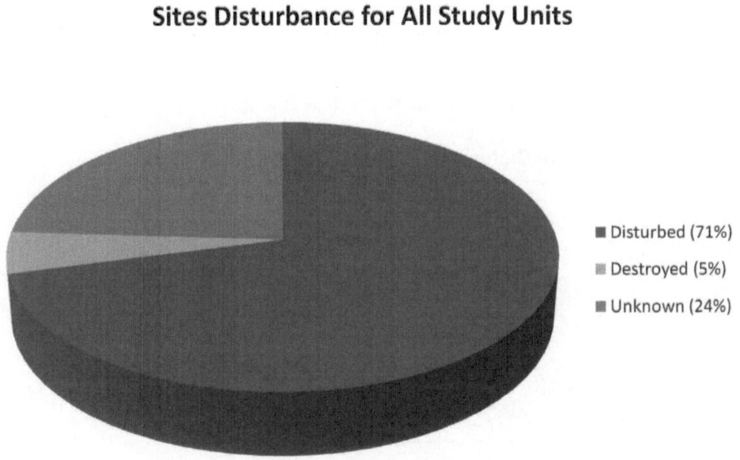

Figure 4.18 Percentage of sites disturbed and destroyed in all *shíshálh* Coast Study Units (n = 129).

Source: Data from British Columbia Archaeology Branch.

of site destruction (Figure 4.20, left). This history is indicated in a variety of ways throughout the study area (Figure 4.21).

In addition, Acheson and Riley (1979) predicted erosion and recreation would after 1977 become the most destructive forces in the region (Figure 4.20, right).

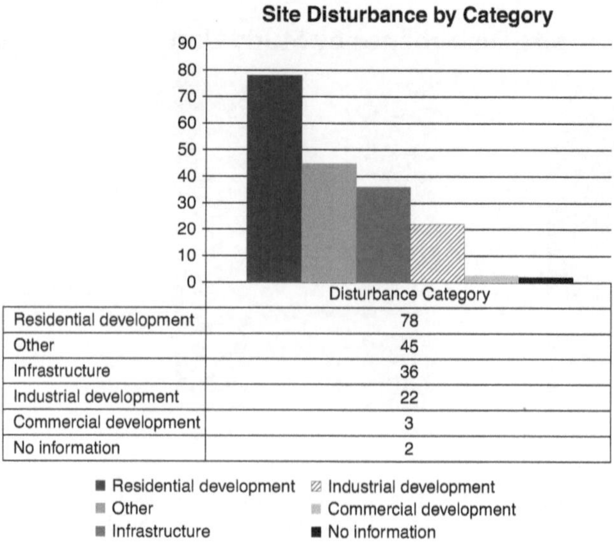

	Disturbance Category
Residential development	78
Other	45
Infrastructure	36
Industrial development	22
Commercial development	3
No information	2

Figure 4.19 Number of site disturbances by category for all *shíshálh* Coast Study Units (n = 129). Today, residential development is the most common cause of site disturbance. The category "Other" includes: archaeological investigation; erosion; recreational use; pot-hunting; dredging; and environmental testing (of these, erosion and recreational were the most significant).

Source: Data from British Columbia Archaeology Branch.

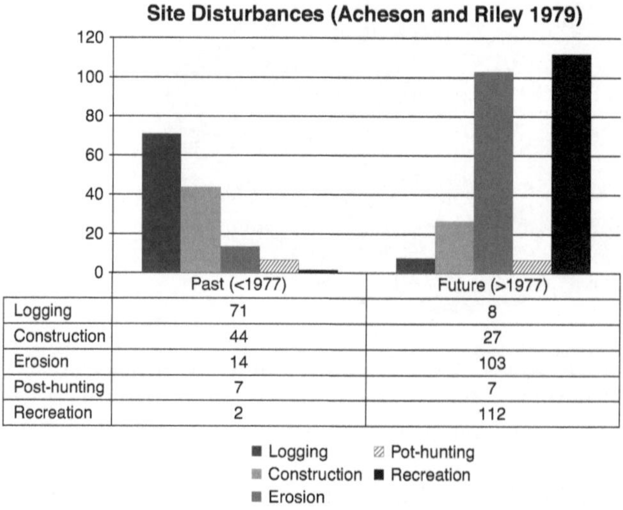

	Past (<1977)	Future (>1977)
Logging	71	8
Construction	44	27
Erosion	14	103
Post-hunting	7	7
Recreation	2	112

Figure 4.20 Number of site disturbances in Acheson and Riley's 1977 northeastern Salish Sea study (n = 186). In 1977, logging was by far the most common cause of site disturbance. They predicted erosion and recreation would become the most destructive forces after 1977.

Source: Data from Acheson and Riley (1979).

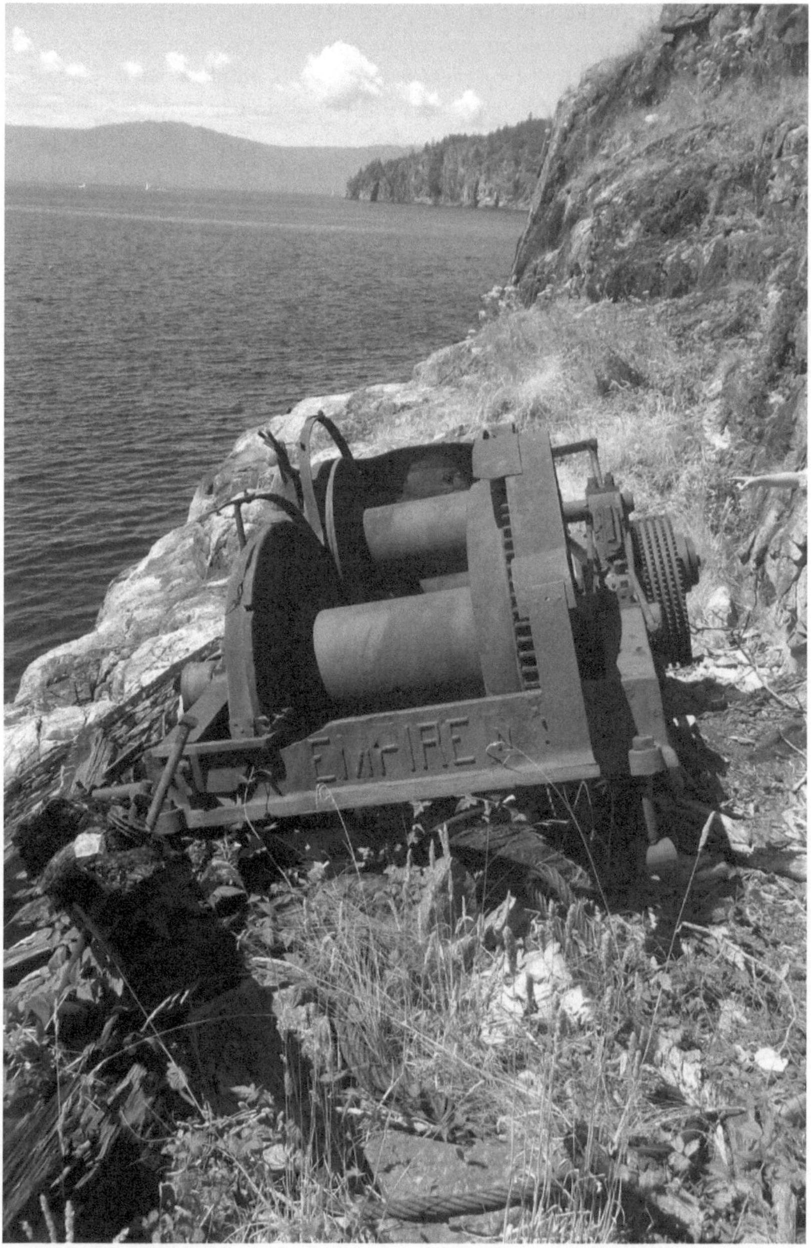

Figure 4.21 Early 20th century "EMPIRE." The northern terminus of my *shíshálh* coast study was on this small bench above the Sakinaw Creek estuary, where I encountered this Vancouver-manufactured donkey engine. Its name is appropriate given that logging was central to empire-building in British Columbia and remains a key aspect of the economy. Historically, logging also had a significant impact on coastal Indigenous heritage landscapes.

In explaining this prediction, they described how "in the last decade there has been a marked shift within this region of the Gulf, towards increased recreational activity *due to an increase in local population*" (1979: 66, emphasis added). Clearly, they perceived a direct link between local population growth and increased impacts.

Based on their calculations and estimates, Acheson and Riley predicted that 99 percent of known sites in their northeastern Salish Sea study area would be threatened by "destructive agencies" in the future (i.e. post-1977). These predictions are strikingly similar to those described by Moss and Erlandson (2008) for coastal Oregon, discussed in Chapter Two.

The *shíshálh* Coast situation is further complicated by Peter Merchant's (2004) findings for Pender Harbour (Figure 4.22). He found that all (100 percent) of the *shíshálh* sites he visited (n = 15) in southern Pender Harbour had been *disturbed*, and, on average, over two-thirds (70 percent) of each site had been *destroyed*. Seven of the 15 sites he visited had been 90 percent destroyed. In addition to characterizing the destructive nature of residential development, Merchant's study raises important questions about the use of the terms "disturbed" versus "destroyed."

To make sense of this language, one must begin by recognizing that the use of the term "disturbed" by archaeologists makes no sense if one takes disturbed to mean mixing of provenience and context. This is because, for scientific archaeologists, a site is destroyed when that site's scientific value is lost. In this way, site disturbance is site destruction. What this means, for example, is that when an

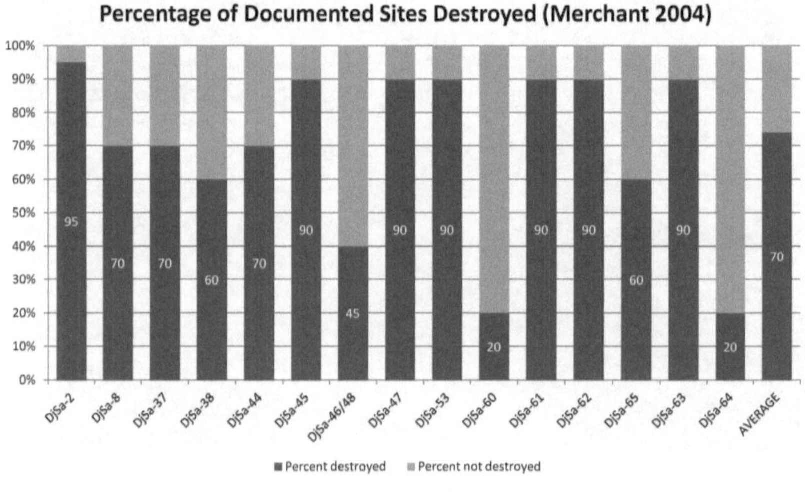

Figure 4.22 Percentage of recorded archaeological sites, identified by Borden Number, destroyed in southern Pender Harbour. On average, 70 percent of every *shíshálh* site that Peter Merchant visited (n = 15) had been destroyed.

Source: Data from Merchant (2004: 53–5).

archaeologist says that a shell midden is 60 percent "disturbed," what they are really saying is that the site is 60 percent destroyed.

Further complicating matters, while 60 percent of that site may be considered "destroyed," the site in its entirety may only be deemed "disturbed" because it does contain 40 percent intact materials. The term "destroyed" has important ethical and legal implications that must be immediately accounted for. As Merchant (2004: 47) points out, regardless of the level of "disturbance," most of the sites he examined are considered by the *shíshálh* Nation to have high cultural significance. The correlation between site destruction and heritage "value" is, therefore, complicated.

On top of all of this, 90 percent of known sites in the *shíshálh* Coast study area are located at less than 15 meters above sea level (Figure 4.23). This 15-meter management zone is designed to (*i*) reflect the uncertainty of future sea level rise (Pilkey and Young 2009); (*ii*) accommodate bank and bluff erosion; and (*iii*) take account of future responses to sea level rise impacts that themselves will negatively affect coastal heritage (e.g. relocating such infrastructure as roads, waterlines, and seawalls). Thus, all landscapes within the 15-meter elevation zone should be considered at-risk. Of more immediate concern is that 88 percent of known sites intersect the contemporary shoreline (Figure 4.24). The key point here is that nine out of ten sites are expected to be directly impacted by sea level rise over the coming decades.

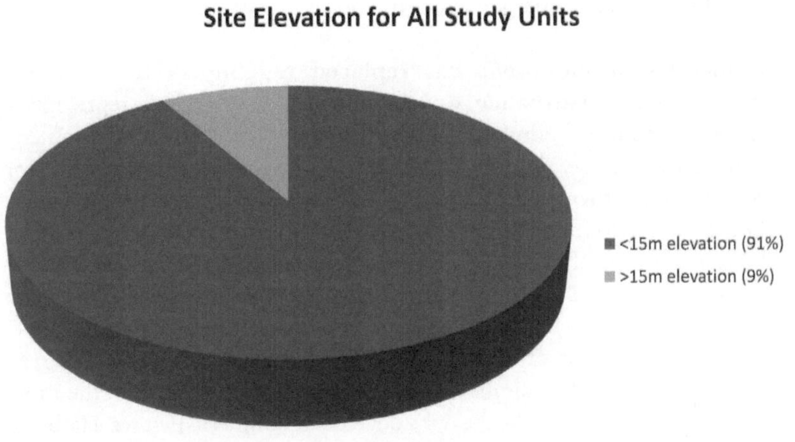

Figure 4.23 Archaeological site elevation (< 15 meters) for all *shíshálh* Coast Study Units (n = 129).

Source: Data from British Columbia Archaeology Branch.

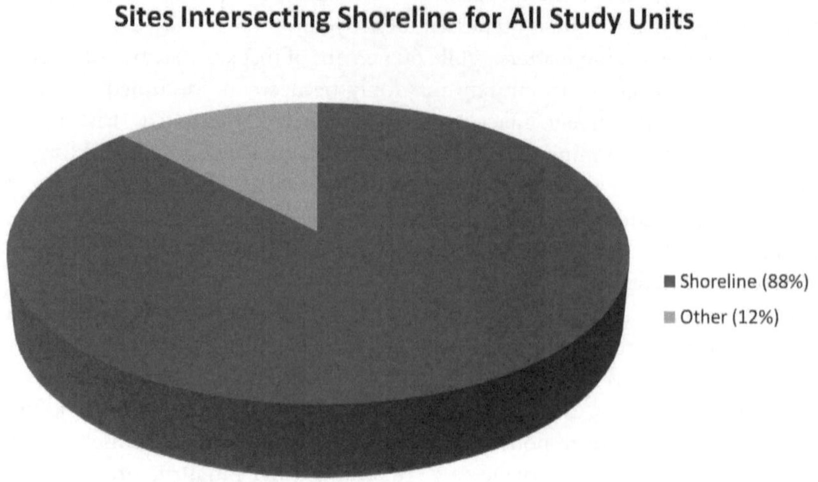

Figure 4.24 Percentage of recorded archaeological sites intersecting the shoreline in all *shíshálh* Coast Study Units (n = 129).

Source: Data from British Columbia Archaeology Branch.

Key Findings

Seven key findings have emerged from my *shíshálh* Coast study:

1 **Residential development has replaced logging as the primary cause of site disturbance on the *shíshálh* Coast.** This represents a shift away from an industrial use of this landscape. This signifies amenity migration or "sea change," characterized by migrants who "move for lifestyle, rather than jobs," choosing destinations with "natural amenity, climate, recreation, and affordable housing" (Gurran 2008: 391). Indeed, about 80 percent of homeowners in the Sunshine Coast Regional District are "non-local" (Figure 4.25). Of those, about 80 percent come from the Vancouver/Lower Mainland area (Figure 4.26).

2 **At least 75 percent of all known sites examined have some degree of disturbance.** This number is ten percentage points higher in the more urban Sechelt area. In the heavily developed southern Pender Harbour, Merchant found that 100 percent of known sites had been disturbed, and, on average, 70 percent of each site had been destroyed.

3 **90 percent of known sites are located at the contemporary shoreline.** All of these landscapes will be severely impacted over the coming decades by residential development and sea level rise. The same percentage of known sites falls within the 15-meter management zone.

4 **For scientific archaeologists, a "disturbed" site is effectively considered destroyed.** This is because such a site has lost its ability to contribute to knowledge about the past.

5 **For the *shíshálh* Nation, all sites are "significant" irrespective of any measure of site destruction or disturbance.** Still, the *shíshálh* management plan seeks to protect heritage landscapes from "disturbance."

6 **While the *shíshálh* have tried to be proactive in the stewardship of their heritage, their ability to effect meaningful change is severely limited by the state and its supporters.** In reality, and despite active participation in the CRM process, the *shíshálh* Nation has very little control over how *shíshálh* heritage landscapes are stewarded.

7 **If the goal of (state-controlled) archaeology/CRM is to protect sites, then the *shíshálh* Coast study demonstrates that it is an ineffective strategy.** The question of whether or not this is the goal will be the subject of the next chapter.

The *shíshálh* Coast study brings awareness to the issue of internal versus external impacts and control. This is illustrated in a 2011 socioeconomic study conducted by the Sunshine Coast Credit Union, which provides important context for my seven findings. The following points come from the report "The Future Business Environment of the Sunshine Coast 2011–2021" (Sunshine Coast Credit Union 2011: 13–14, 34–5, 52):

- Population growth driven mainly by migration largely from Vancouver and surroundings, two-thirds of net migration is in age group 45 years or older;
- Population aged 55+ increased more than 140 percent since 1986 compared to less than 45 percent in the under 55 age groups;

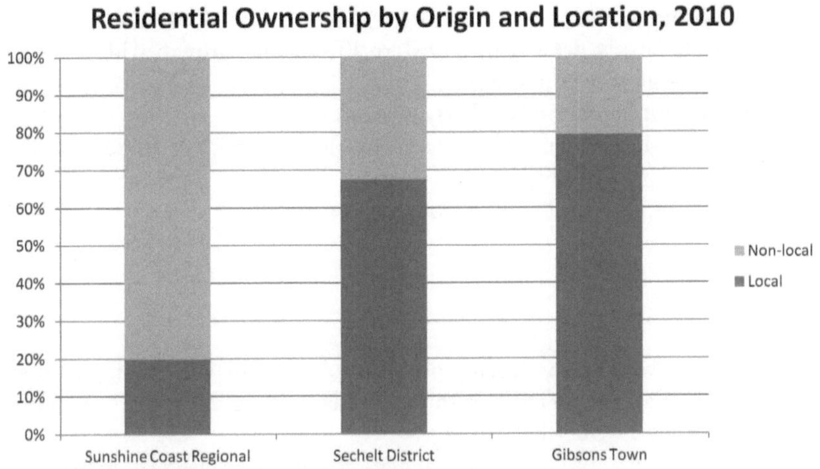

Figure 4.25 Residential ownership by origin (local vs. non-local) and location (Sunshine Coast Regional District, Sechelt, and Gibsons) for the year 2010. In the Sunshine Coast Regional District (i.e. everything other than the towns of Gibsons and Sechelt), 80 percent of residences are owned by non-locals, compared to 20 percent in the Town of Gibsons.

Source: Data from Sunshine Coast Credit Union (2011: 45).

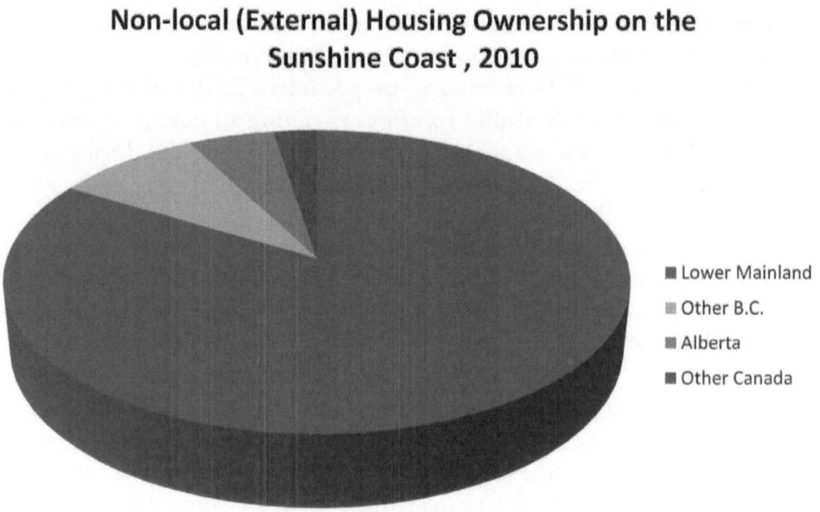

Figure 4.26 Non-local housing ownership on the Sunshine Coast by origin, 2010. About 80 percent of all non-local Canadian homeowners on the Sunshine Coast derive from the Lower Mainland (i.e. City of Vancouver and surroundings).

Source: Data from Sunshine Coast Credit Union (2011: 46).

- Aging of the post-war baby boom and the older age make-up of migration to the Sunshine Coast are main factors;
- Economy less dependent on natural resources, more service-industry-oriented economic base;
- Economy largely dependent on external factors for growth with some lift from local growth conditions;
- A significant number of workers commute;
- Large non-local ownership of residential properties; generates seasonal increase in population; and
- Majority of non-local owners from lower mainland.

The study ends with this "forecast summary" (Sunshine Coast Credit Union 2011: 57–8):

- Higher growth largely depends on stronger economy and housing markets in Metro Vancouver and British Columbia, leading to more in-migration and non-local ownership activity.
  - Higher in-migration and population growth drives real estate, construction, and retail employment.
- Non-local residential ownership continues to increase.
  - Age structure of population shifts towards older age groups; by 2021, more than 50 percent of resident population aged over 55 and 33 percent aged 65 years or older.

The report's authors conclude that in order to "retain, expand, and attract businesses and people," efforts need to be made to "[p]romote and market [the] Sunshine Coast."

Summary

Sea level rise-induced flooding, exacerbated by increased storminess, will have a devastating impact on the vast majority of *shíshálh* heritage landscapes. Further, responses to sea level rise—notably the inevitable landward/upward relocation of infrastructure such as roads, structures, and seawalls—will exacerbate the problem (see BC 2013b: 7, Figure 3; Pilkey and Young 2009: 179). As shown in my *shíshálh* Coast study, archaeology/CRM has been ineffective at protecting heritage sites from development; there is no reason to expect, therefore, that it will be effective at protecting those landscapes from rising seas.

It is primarily through amenity migration that contemporary *shíshálh* maritime heritage destruction is linked to the core that is Vancouver. In this regard, the situation is no different from the previous period when logging was the dominant destructive force. As such, there has been no change in the dominance of the core over the periphery in issues of local control over heritage. In the next chapter, I discuss why archaeology/CRM persists as the mainstream response to coastal change.

5 Problematizing the Heritage Crisis

> It is only through fundamental change at the center of the system, from which the pressures on the planet principally emanate, that there is any genuine possibility of avoiding ultimate ecological destruction.
>
> (Foster 2009: 276)

The *shíshálh* Coast study shows archaeology/cultural resource management (CRM) to be ineffective at protecting Indigenous heritage landscapes, and highlights how heritage destruction principally emanates from the core. In the past, the primary cause of *shíshálh* heritage destruction and degradation was industrial resource extraction. Today, it is amenity migration and residential development. In the future, rising seas and responses to coastal flooding will consume much of what remains of the physical elements of *shíshálh* maritime heritage landscape. My goal for this chapter is to better understand how and why things have gone so terribly wrong.

Towards that goal, I shift the discussion away from the *shíshálh* Coast and toward the center of power, in this case Vancouver–Victoria. This pairing is important because while the impacts on the coast principally emerge from the economic center (Vancouver), it is the legal-political thus bureaucratic center (Victoria) that "permits" the development. In addition, both Vancouver and Victoria house the major universities that discipline archaeologists in resourcism and management, the two key elements of compliance archaeology. Further, the vast majority of professional archaeologists are permanent residents of these cores, thus they naturally take on the urban or metropolitan worldview.

I apply the "power-elite dynamic" to the *shíshálh* Coast study, deconstructing the concepts that are "management" and "resourcism." I then show how archaeology/CRM is a form of neoliberal statecraft, a government technology designed not to save heritage from development, but rather to clear it from the landscape. A major challenge to seeing how archaeology/CRM works in service of the state is that archaeology constitutes an ideology; thus, its position at the center of the system represents a "taken-for-granted," rendering it largely "invisible," especially to its practitioners. A similar challenge is the slow pace at which the maritime heritage crisis typically unfolds, a tempo that also works to obscure archaeology's role in it. I end with a "problematique," an attempt to (re)connect the maritime heritage crisis to broader global processes and problems.

The Power-Elite Dynamic—Scale, Power, and Land Use

> The beliefs underlying capitalism are the most important ideological cause of contemporary adverse environmental changes because capitalism drives the scale changes that have most intensified human pressures on the environment.
>
> (Bodley 2008b: 53)

> A great deal of archaeology is ideological practice, practice which sustains and justifies a capitalist present.
>
> (Shanks and Tilley 1987: 62)

Anthropologist John Bodley's *power-elite hypothesis* states that growth in the scale of sociocultural systems is an elite-directed process that concentrates social power (Bodley 1999, 2001, 2002a, 2002b, 2003). It predicts that when power is commercially organized and the scale of a community grows (urbanizes), "economic power becomes more concentrated" (1999: 595). In a nutshell, "big may not be better for everyone" (2002b: 73).

By way of example, Bodley's (1999: 595) study of contemporary communities in eastern Washington State shows that:

> Elite property owners, who most benefit from growth, assume a larger role in municipal government, where they can encourage further growth through municipal annexations and zoning changes. Thus, as elite power becomes increasingly concentrated, the growth process itself tends to become self-perpetuating. In the [Washington State] example, small, no-growth municipalities appear to be politically more democratic than larger-scale, growing municipalities and household well-being in them more equitably distributed.

Population pressure is the most widely accepted anthropological explanation for intensified environmental problems such as land degradation (2002a: 166), and "the most striking general conclusion to be drawn from the cultural ecological data in the anthropological record is that the speed and scale of resource depletion and environmental degradation accelerates with increases in the scale of culture and the concentration of social power" (2002a: 163). The consequence of the power-elite dynamic in a commercial scale culture (e.g. British Columbia) is increasingly centralized social power in the core and heritage destruction in the periphery (Bodley 2008a, 2008b).

In this framework, archaeology is not just simply a product of the power-elite dynamic, but it is a tool of the power-elite. For example, in terms of Indigenous heritage management in British Columbia, Vancouver–Victoria controls all the key aspects of the process, from governance (Dent 2012) to ideology production (Nicholas and Hollowell 2007). In the *shíshálh* Coast study, heritage destruction is largely associated with amenity migration, the result of people trying to "escape" an increasingly toxic and nature-less, thus sick, urban environment (La Salle 2014a). The outcome of urban growth for "peripheral" Indigenous cultural heritage is demonstrated clearly in the number of sites disturbed and destroyed as a result of out-migration from the Lower Mainland to the amenity-rich "Sunshine Coast."

In terms of the present inquiry, the outstanding question, then, is what specific roles do scale and power play in resource management? Before delving into that question, it is useful to pause and think about the very much taken for granted ideas/ideologies that are management and resourcism. My premise here is that the best way to understand "resource management" is through its component parts. After considering each, I examine them in the context of the power-elite dynamic.

Cultural Resource Management (CRM) as Management

> **management** the process of planning, organizing, coordinating, and directing the productive process in an economic enterprise.
>
> The term is also applied to those persons, organized in a hierarchy, who carry out the management process, that is, who perform these tasks or functions.
>
> (Theodorson and Theodorson 1969: 240)

The verb "manage" is derived from the Latin *manus*, meaning "hand," and later the Italian word *maneggiare*, meaning "to handle," especially in regards to tools and weapons. In its current usage it refers variously to: (*i*) the action of controlling or manipulating a person, animal, etc.; the exercise of control; (*ii*) to control (a person or animal); to exert one's authority or rule over; (*iii*) to take charge of, control, or direct (a household, institution, business, state, etc.); (*iv*) to organize the activities of (a person or group of people); (*vi*) formerly, to cultivate, till (land); later, to maintain and control (the environment, an area, forest, nature reserve, etc.); also, to conserve natural resources such as game, fish, timber, wildlife, etc.; and (*vii*) to manipulate or tamper with in order to deceive; to adulterate, doctor. To "manage," therefore, is to exert control, the latter being a synonym of power.

The noun "manager" refers variously to: (*i*) a person who organizes, directs, or plots something; a person who regulates or deploys resources; the wielder of a weapon; (*ii*) a person skilled in controlling or regulating finances and expenditure, especially of a household; a person who runs a household (well or badly); and (*iii*) a person who manages (a department of) a business, organization, institution, etc.; a person with an executive or supervisory function within an organization, etc. "Managers," therefore, are individuals or groups who control some "thing," be it tangible or intangible. In addition, the latter definition (*iv*) hints at the bureaucratic and financial nature of modern management.

The noun "management" refers variously to: (*i*) organization, supervision, or direction; the application of skill or care in the manipulation, use, treatment, or control (of a thing or person), or in the conduct of something; (*ii*) cunning, manipulation, trickery; the use of scheming, intrigue, prudence, etc., to achieve something; (*iii*) a negotiation (to be in a management); to be engaged in negotiations; (*iv*) a governing body of an organization or business, regarded collectively; the group of employees which administers and controls a business or industry, as opposed to the labor force; and (*v*) the responsibility for and control of the resources of a company, department, or other organization. As indicated by Theodorson and Theodorson's (1969) definition above, there is an essential economic aspect to management.

Management, therefore, refers to the collective work that a manager does and managers do. In *Economy and Society*, Max Weber (1964 [1922]: 956–8) identified six bureaucratic values that he saw as essential for good management. Those are: (*i*) imperial positions (top-down control); (*ii*) chain of command (hierarchy of power); (*iii*) rule-governed decision making (rules of conduct regarding chain of command); (*iv*) defined responsibilities (specialization); (*v*) bounded authority (rules of conduct for workers); and (*vi*) professionalism (pursuit of excellence). These values characterize the CRM system today (Hutchings and La Salle 2015a, 2015b). Archaeologists who operate in this system—referred to as "archaeobureaucrats" by Joshua Dent (2012)—"are paid to assess projects, apply for permits, carry out fieldwork, write technical reports, and in effect 'manage' resources" (Lyons et al. 2012: 8). According to Weber (1964 [1922]: 225), bureaucratic administration means, fundamentally, "domination through knowledge."

Today, scholars see the highly bureaucratic resource management institution (see King 2009) as containing the "seeds of failure" (Acheson 2006: 124). Ken Lertzman (2009: 347) reminds us that there is "nothing inherent in the idea of 'management' that implies an objective of sustainability or conservation." Indeed, if management:

> embodies values and norms determined by its social context, then to the extent that they follow social norms, the goals of a management system may legitimately not include a 'conservation' ethic. If a society values commodity production over conservation, then we will see management systems develop that reflect this.
>
> (Lertzman 2009: 347)

Sociologist Sylvia Hale (1990: 518–19) describes the major design flaw this way: "The loyalty of officials lies not with the general public or the electorate, but with the bureaucracy itself. Their vocation is to serve their official duties." Employees within the bureaucracy are "even more rigidly subject to its regulations. They operate as cogs in the machine." Hale suggests that the "major requirement for their position is unquestioning and strict adherence to written regulations within their narrowly defined areas of jurisdiction. Their individuality has no place within such a system, for it would disrupt the calculated order."

Indeed, resource managers have very little control over the nature of their work. In part, this is due to the institution's hierarchical, bureaucratic structure, and in part because most managers have been socialized into the cultural tradition of resource management, which is hierarchical, bureaucratic, and scientific. The first helps explain why resource managers so often carry out what they personally see as counterproductive policies, the second why so many resource managers believe in the system to begin with. The first makes resource managers part of the problem; the second makes them blind to that fact.

Resource managers do not conserve things so much as they help carry out the elite-driven project that is growth, development, and progress (Hutchings and La Salle 2015a). In his study of institutional failure in resource management, James Acheson (2006: 125) considers imperial or top-down management, where powerful agencies "ride roughshod" over local peoples and landscapes. The *shíshálh* Coast study demonstrates exactly this conflict between Indigenous interests in heritage stewardship and the government-permitted management of archaeological sites. These two concepts—stewardship and management—lie in contrast (Lertzman 2009: 347): "While managers can 'manage' to achieve whatever objectives society has given them, stewards always have an obligation to the resources or ecosystems themselves. The idea of stewardship is very consonant with traditional resource management principles."

CRM as Resourcism

> Since by "resources" we generally mean anything which can be bent to human use or consumption, then virtually anything is potentially a resource. … In Heidegger's earlier terminology, it is the view of the world as a "standing reserve" awaiting exploitation by human societies.
>
> (Evernden 1995: 563)

Resource managers embody the values and norms determined by their social context, which is the commercial-scale culture of consumption. It seems, then, that resource management's governing paradigm centers on the commodification of resources; thus, it also includes aspects of bureaucratization, specialization, and globalization.

This philosophy is captured in *resourcism*, which involves the conversion of heritage into commodities. Through this conversion, "resources" are produced that are in need of "management." Evernden (1995: 563) addresses the issue this way:

> while it is commonplace to speak of resources (and nowadays even "human resources"), the striking feature of resourcism is its tacit expectation that we exist solely in a domain of resources, whether of present use or potential use.

> … To its critics the disturbing feature of resourcism is the single, homogenizing feature of value that it brings to its worldview. … Resourcism constitutes a kind of "lowest common denominator" of material existence and makes it difficult or impossible to speak of heterogeneity of values or to deal in anything except economic evaluation.

Resourcism is an *ideology*, meaning it comprises a "system of values that justify certain kinds of actions or ways of life" (Hale 1990: 568). It is the capacity of the dominant class to "rule through control over prevailing ideas or cultures. It ensures that the mass of people accept as legitimate the activities that directly benefit the dominant class" (1990: 568).

The resourcist ideology, produced by the power-elite dynamic, is *hegemonic*; that is, the power-elite supply the values, sense of heritage, meaning, and motivation for managing "resources" that are in line with their own capitalist interests. In academic archaeology, this is achieved through a focus on science, progress, and materiality. In CRM, as in other management areas, resourcism permits and promotes an agenda of growth, development, and scientific progress. Together, these ideologies are responsible for the depletion of the resource.

Addressing this issue, Thomas King (2009: 7) accuses the American resource management system of "whitewashing" heritage destruction:

> we now have bureaucracies overseeing environmental impact assessment (EIA) and cultural resource management (CRM), and we have well-heeled private companies doing EIA and CRM work under contract. What we do not have is an orderly system for actually, honestly considering and trying to reduce impacts on our natural and cultural heritage. It's all pretty much a sham. … we've let things develop in such a way that we're spending lots of money on mere public relations efforts, slathering the lipstick on the pigs of ill-considered development. We ought to change our system.

In sum, the project that is growth, development, and progress is elite-directed and the institution of archaeology/CRM is one technology that facilitates that project. By converting heritage landscapes into archaeological resources, state-sanctioned managers can control heritage from afar and in service of the state. The driving force behind this project is capitalism.

Heritage experts have discussed archaeology as a form of statecraft (Fowler 1987; Smith 2004; Trigger 1989), and archaeology's capitalist and colonialist tendencies have been queried and clarified (Hamilakis and Duke 2007; McNiven and Russell 2005; Shanks and Tilley 1987; Zorzin 2014, 2015). Archaeology's ties to neoliberalism, however, have received less attention. An extension or form of capitalism, neoliberalism emphasizes free market relations (Springer 2010: 1025). As an ideologically hegemonic project, *neoliberalism* maintains that "elite groups, organized around transnational class-based alliances, have the capacity to project and circulate a coherent program of interpretations of the world on to others" (2010: 1032). In addition to class power, Simon Springer

(2010) highlights two other key facets of neoliberalism: its bureaucratic formation and its political formation.

In this context, archaeology operates as a technology of government (Smith 2001, 2004), used to produce and mobilize heritage in support of neoliberal state interests (Coombe 2012). On the Sunshine Coast, this power-elite dynamic is manifested through "external" development—that is, "development" conceived of by the core and carried out in the periphery. CRM serves these interests by facilitating and "permitting" development, rather than by protecting heritage from those interests. Towards that end, CRM may be seen not as "failing" at conservation, but rather as succeeding at its true function.

State Control of Indigenous Heritage

> Who controls the past controls the future: who controls the present controls the past.
>
> (Orwell 1949: 248)

> What could be more reasonable than a desire to ensure that you are the custodian of your own cultural heritage? And what could be more unreasonable than holding another people's cultural heritage, of ongoing significance to them, in your hands?
>
> (Asch 2009: 394)

Site destruction is a major concern for descendant communities. In light of my *shíshálh* Coast study, a critical question emerges: Given that CRM does not seem to prevent heritage site destruction, and that this appears well known to archaeologists working in the province and elsewhere, why does society continue to pursue these ineffective measures with so much determination?

George Nicholas and Julie Hollowell (2007: 2) describe how, historically, "archaeology has served the needs of the nation-state" as it "disenfranchises [descendant communities] from their own histories." Eldon Yellowhorn views the matter this way (pers. comm. 2014):

> CRM is statecraft that facilitates the statutory expropriation of Aboriginal rights, because if culture is a resource then its constitutional status is clear. The archaeological record stops being the labour of the ancestors of Aboriginal people and instead becomes analogous to concretions that form naturally in the land. On the other hand, archaeologists become the people who go into an area to harvest the cultural resources before corporate entities harvest the natural resources.

John Welch and Neal Ferris (2014: 94) draw attention to "inconsistencies between what archaeologists say archaeology is doing (i.e., conserving the archaeological record) and the consequences that the unprecedented rate and scale of harvesting the record is having, primarily within CRM practice."

These concerns have been expressed by many First Nations across British Columbia. A number of critical points were raised by late Musqueam Chief Ernest Campbell in an open letter sent to various levels of government, Indigenous groups, and the United Nations (UN) (Campbell 2012; see also Hutchings 2012a). Chief Campbell (2012: 4) suggested that British Columbia's *Heritage Conservation Act* (HCA) is:

> being implemented so as to destroy historical First Nations sites so development can take place. It places little value on pre-colonial history or First Nations wishes. To the best of our knowledge, it has never prevented a commercial development despite generations of First Nation protest. Now it is being used to legalize the destruction and desecration of one of the largest and most significant remaining archaeological sites or Middens in Canada for a few condominiums that could be built elsewhere.
>
> (Campbell 2012: 4)

In pursuit of solutions to this, Eric McLay et al. (2008: 150) explore Central Coast Salish Hul'qumi'num customary laws and concerns about the protection of their archaeological heritage, calling its destruction by modern development a "chronic problem." Hul'qumi'num elders assert that "this loss of and disrespect shown towards their archaeological heritage is contrary to many of their cultural teachings." This includes CRM:

> We're concerned about archaeologists operating in our territory, as well as other First Nations territories, archaeologists who are lacking in the kind of ethics we require. We see instances, over and over again, of consultants being employed who are not looking for evidence of First Nations use and occupation but are looking for evidence of a pay cheque.
>
> (Cowichan Elder Joey Caro, quoted in McLay et al. 2008: 182)

Elders fear that, "in continuing to bend their customary laws to permit the removal and reburial of ancient human remains that are threatened by development projects, they will ultimately undermine their own laws against disturbance and, ultimately, leave these sites vulnerable to exploitation by developers" (2008: 178).

CRM, representing the practice of archaeology today, is thus viewed with trepidation by the descendants of the heritage that is being "managed." As such, and like coastal resource managers, cultural resource managers, acting as specialists, are increasingly focusing on integration and co-management, what is referred to as "consultation" and "collaboration" (Atalay et al. 2014; La Salle 2010, 2013, 2014b; La Salle and Hutchings 2015; Merchant 2010).

These approaches assume that the system itself is not the problem. However, as Smith (2004: 196) describes, archaeology has been viewed by the government to be useful:

> in so far as it had a concrete policy effect, because of the social claims made by Indigenous peoples in the 1960s and 1970s. Archaeology as an area of expertise could be mobilized to help the state make sense of, and "govern," Indigenous claims.

Archaeology/CRM thus plays a part in 'governing' populations and representations of their pasts" (Smith 2004: 197). Smith defines archaeology/CRM as a "technology of government" (Mallette 2009) or "ideological state apparatus" (Althusser 1971), designed to solve what is still considered to be "the Indian Problem." The social impacts of state management of Indigenous heritage are considered in the next section.

Archaeology as Slow Violence

> As Marx showed so convincingly in the closing chapters of the first volume of *Capital*, the primary accumulation of capital was effected through violence and plunder, and the same methods continue in daily use throughout capitalism's dependent colonies and semi-colonies.
>
> (Baran and Sweezy 1966: 336)

> Whereas individual people can become dislocated by misfortunes in any society, including tribal, feudal, and socialist ones, and whereas the downfall of any society produces mass dislocation, only free-market society produces mass dislocation as part of its normal functioning.
>
> (Alexander 2008: 60)

The outcome of state archaeology/CRM for Indigenous peoples is violence. Archaeological violence is manifested on the ground through what psychologist Bruce Alexander (2008) calls "dislocation" and Glenn Albrecht (2005) terms "solastalgia" (Hutchings 2014). Working in a poverty-stricken neighborhood with a high Indigenous population in urban Vancouver, British Columbia, Alexander demonstrates how neoliberal policies work to "dislocate" First Nations individuals from their past (heritage, place, identity) and present (community, place, identity), often resulting in addiction, homelessness, and violence. Alexander (2008: 15–16) considers dislocation in historical and economic context:

> Although carried out with relatively little lethal violence, the dislocation imposed upon aboriginal people in British Columbia was arguably more severe than in other Canadian provinces. British Columbia's Indians were dislocated physically from their land, socially from their culture and families, linguistically from their native tongues, economically from their livelihoods, and spiritually from their ancestors, and gods.

Environmental philosopher Glenn Albrecht et al. (2007: 45) use the term *solastalgia* to characterize the distress caused by dislocation from one's environment, which includes pain, loss, and a feeling of being "undermined by forces that destroy the potential for solace to be derived from the present." Albrecht (2005: 46) connects solastalgia to "more serious health and medical problems such as drug abuse, physical illness and mental illness (depression, suicide)." Indigenous people are particularly vulnerable to these health consequences because of their strong connections to place and their experience of ongoing colonization (2005: 54), which has resulted in the loss of control over their heritage and identity. Sutton et al. (2013: 7–8) describe the impacts of solastalgia in the context of CRM taking place in a major coal mining district in Australia. They acknowledge "Indigenous attachment to the environment as an intimate spiritual connection," and warn "the impacts of coal mining therefore not only destroy that environment but irrevocably damage people's connections to country."

One problem in identifying solastalgia in the context of heritage destruction is that it is a "slow" process, thus making it difficult to see (Nixon 2006–2007, 2011a, 2011b). As Rob Nixon articulates:

> We are accustomed to conceiving violence as immediate and explosive, erupting into instant, concentrated visibility. But we need to revisit our assumptions and consider the relative invisibility of slow violence. I mean a violence that is neither spectacular nor instantaneous but instead incremental, whose calamitous repercussions are postponed for years or decades or centuries. I want, then, to complicate conventional perceptions of violence as a highly visible act that is newsworthy because it is focused around an event, bounded by time, and aimed at a specific body of bodies.
>
> (2011b: n.p.)

Heritage is fundamental to cultural survival (UN 2008). Archaeology/CRM, as the process by which Indigenous heritage places are destroyed for capitalist expansion, constitutes violence consistent with the genocidal policies of the colonial governments (Hall et al. 2009; Regan 2010). However, archaeology/CRM is slow violence because the "loss" of heritage sites, or rather their permitted destruction, is incremental and rarely newsworthy (see, however, Hutchings 2011, 2012a; Mapes 2009; Stapp and Longenecker 2009).

In archaeology, the violence produced through CRM constitutes an externality—something "outside" of the institution, thus beyond the control and concern of individual archaeologists (Hutchings and La Salle 2015a). As such, neoliberal violence is accepted unquestionably as part of the permanent structure within which the modern heritage manager must operate. In reproducing the structure, archaeobureaucrats whitewash their role in the Indigenous heritage crisis.

Yet, archaeologists represent the frontline of Indigenous heritage landscape destruction—they are the ones that ultimately "authorize" and make "professional" archaeology/CRM. It is their academic currency that gives the project legitimacy. They are the ones that "permit" (make legal) the clearance of heritage landscapes. Archeologists are not benign actors in this neoliberal state project; they are largely true believers in their cause, and act accordingly. This view comports with the new psychology of tyranny (Haslam and Reicher 2012; Reicher and Haslam 2006), whereby *tyranny* constitutes "an unequal social system involving the arbitrary or oppressive use of power by one group or its agents over another" (Reicher and Haslam 2006: 3). At issue here is the degree to which archaeologists are aware of their complicity in the violence inhered in their practice.

In psychologist Stanley Milgram's famous Obedience to Authority research, volunteers were directed to electrocute others in the name of scientific discovery—in some cases, to the point of death. They did so because they believed their work was "'service' in the cause of 'goodness'" (Haslam and Reicher 2012: 1), contributing to human progress. Similarly, although archaeology/CRM leads to the destruction of Indigenous heritage landscapes, archaeologists, "rather than being distressed by their actions," construe their practice as "saving the past," not destroying it. Like Milgram's test subjects, archaeologists participate in tyranny and violence because of their unwavering belief in science and progress.

Problematique

> Resource management is at a crossroads. Problems are complex, values are in dispute, facts are uncertain, and predictions are possible only in a limited sense. The scientific system that underlies resource management is facing a crisis of confidence in legitimacy and power. Top-down resource management does not work for a multitude of reasons, and the era of expert-knows-best decision making is all but over.
>
> (Berkes et al. 2007: 308)

Building upon observations made in the preceding chapters, 11 theses are presented here that define my "problematique." As a theoretical device, the problematique first appeared in 1970 in the Club of Rome's (1970: 12–13) famous "Predicament of Mankind" study. There, the concept was introduced as follows:

> When we consider the truly critical issues of our time such as environmental deterioration, poverty, endemic ill-health, urban blight, criminality, etc., we find it virtually impossible to view them as problems that exist in isolation—or, as problems capable of being solved in their own terms. For even the most cursory examination will at least reveal the more obvious (though not necessarily the most important) links between problems. Where endemic ill health exists, poverty cannot easily be divorced from it, or vice versa. … [I]f we try to solve any such problems exclusively in their own terms we quickly discover that what we take to be the solution of one category of problem may itself generate problems of another category.

Towards that end, the Club of Rome compiled a 49-point problematique of "Critical Continuous Problems" (1970: 14–16), included here as an Appendix. My 11 theses must, therefore, be considered in addition to and alongside that list.

1 Coastal development has had and continues to have a tremendous impact on Indigenous maritime heritage.

Data from the Salish Sea demonstrates the enormous effect that a Western ideology of growth, development, and progress has had on maritime heritage. For Indigenous descendent communities, this means the physical destruction/desecration of heritage sites and landscapes. Importantly, such offenses are part and parcel of a long list of other very serious violations (Culhane 1998; Harris 1997; Regan 2010). Today, "negative impacts" are normal operating procedure in CRM (McLay 2011). They are merely the "cost of doing business."

For the most part, non-Indigenous heritage falls outside the practice of CRM in British Columbia. Conversely, settler heritage falls under the purview of other government agencies (often related to urban revitalization and parks), regional/local historical societies and organizations, mass media, and educational textbook publishers. The heritage celebrated largely focuses on the history of discovery, conquest, and capitalist exploitation of a virgin or untamed "wilderness." Small local museums center specifically on the establishment of an industrial maritime-based economy focused on logging and fishing, along with pioneer settlement, growth, and expansion. Collectively, these stories comprise what I call the "Northwest Coast settler narrative," part of the larger endeavor to manufacture national identity (Anderson 1983). As such, the heritage celebrated in mainstream society is the heritage of coastal growth, development, and progress; the heritage that suffers this narrative is often Indigenous.

2 In British Columbia, heritage specialists should plan for a sea level rise of 1 meter by 2100, 2 meters by 2200, and 3 meters by 2300, a process that will have a catastrophic impact on maritime heritage landscapes.

However, sea level rise is but one threat among many when it comes to climate change. Impacts on natural heritage cannot be separated from those acting on cultural heritage landscapes; the loss of one means the loss of the other. There is significant uncertainty in the predicted rate of sea level rise, globally and locally. However, all coastlines will be impacted in some way. These impacts are and will be compounding the long history of more direct development impacts. It is critical to recognize that sea level rise is itself a product of industrial development. Consequently, it should be treated as such.

3 Taken together, coastal development and sea level rise represent a modern, concatenated "maritime heritage crisis."

Until very recently, the problem has constituted a "quiet crisis" as it has received so little attention. Indeed, archaeologists and cultural heritage specialists have been late on the climate change scene despite the preservation and conservation of archaeological sites usually being touted as the primary concern by this community. In CRM, this is highlighted by King (2009: 166), who devotes the final pages of *Whitewashing* to the subject:

> I've often paused in writing this book, sometimes for hours, sometimes for months, overcome by the feeling that I'm fiddling while Rome burns. What earthly point is there, I ask myself, in diddling about with the adjustments in the way we consider environmental impacts—and impacts on things like old buildings and archaeological sites, for heaven's sake—when the atmosphere's warming, the polar ice is melting, the seas are rising? Doing field work as I sometimes do on a Pacific atoll, I can see the water rising, and I doubt if anything is going to save a host of island nations and a lot of continental waterfront from inundation. The costs of this disaster in terms of natural and cultural heritage—including living, functioning human populations and their social systems—will be far beyond anything we can imagine measuring or mitigating.

As King (2009: 167) observes, the impacts of coastal change responses "may be greater in some ways than the impacts of climate change itself. And unlike climate change these impacts will be subject to human control, human influence."

4 The mainstream response to the maritime heritage crisis is CRM.

Predicated on resourcism, archaeology/CRM privileges the role of physical heritage and sites over intangible meanings and landscapes. Mainstream strategies typically revolve around the acquisition of more data through methods that include coastal surveys, site valuation, and proposed mitigation strategies. This is reactive and has little to do with heritage preservation.

Indeed, the Union of British Columbia Indian Chiefs (UBCIC 2013: 13) note how the values espoused by the Archaeology Branch and its "desire to 'assist the development industry' are in stark contrast to how First Nations value heritage sites."

> the provincial Archaeology Branch sees heritage as archaeological; therefore a site that does not have a physical or archaeological expression, or human alteration of the landscape, is usually not recognized as a provincial heritage site. This has made it virtually impossible for First Nations to rely on this legislation to help protect our spiritual, sacred and traditional use sites—all of which frequently have little physical or archaeological evidence.
>
> (UBCIC 2013: 15–16)

As such, CRM is fundamentally about compromise and does not account for the fact that archaeology might be part of the problem.

5 Archaeology/CRM emphasizes a scientific understanding of "the past" and the material record, notably the archaeological "site."

Archaeologists helped construct British Columbia's CRM laws. Today they dominate the workforce that sustains CRM, and they hold top government positions that oversee the industry.

This mainstream scientific approach, concerned with the "material record," is narrow in scope and relevance. It does not deal with either the intangible or the present; thus, its relevance to contemporary society and their heritage is questionable.

6 *Because of their connection to the state, archaeologists have a monopoly over the production of heritage meanings; the public and descendant populations have little direct involvement in the process.*

North American archaeologists, operating as specialists, have come to dominate the production of knowledge about Indigenous pasts; yet, Indigenous views on their own past have been routinely ignored (Markey 2001). Indeed, while the permitting process in British Columbia requires a notice of intended archaeological work be sent to affected First Nations communities, in practice they have little control over the process, which may well proceed without their involvement. This point has been made clear by King (2009) for the United States and raised by the UBCIC (2013) for British Columbia, as well as by the *shíshálh* Nation. The result has been that the histories produced through archaeology bear little resemblance to the cultural understandings of descendent communities (Yellowhorn 2002).

7 *In British Columbia, CRM is fundamentally about managing Indigenous heritage, populations, and rights.*

Laurajane Smith (2004) links the rise of CRM with the Indigenous rights movement, reinforcing how controlling heritage is about controlling people. As Michael Asch (2009: 395–6) points out:

> The state of Canada—a state that did not even exist when much that is cultural heritage of First Nations was created, largely does not respect the names given to places before the arrival of European settlers, and established the rule of law over First Nations on the basis of a colonial ideology invented at the time of the British imperium—is considered to hold legitimate authority over First Nations cultural heritage. … [I]n British Columbia today, First Nations still have to appeal to the courts, government officials, or boards regulated by the province in order to persuade people to respect gravesites when developing lands that are part of their cultural heritage.

In this view, archaeology/CRM may be considered a technology of the state, mobilized to minimize resistance to the national and imperialist project that is growth, development, and progress. CRM can, therefore, be considered a token gesture or "trade bead" in the pursuit of profit, representing not a departure from but the continuation of colonialist policy. Indeed, while some individuals and groups have "moderated their thinking and practice with regard to these issues, the state and its institutions remain firmly in control" (Asch 2009: 396).

8 Consultation and collaboration are not solutions to any of these problems, for they largely reproduce the current socioeconomic structure.

Consultation is not a solution within the current system "because it is being defined and implemented within a wider capitalist and colonial framework" that "does not allow for devolution of power and authority from the state to [sovereign] peoples" (Cobb 2010: 2). As Miranda Cobb (2010: 2) puts it, "as long as no broader steps towards decolonization are taking place, meaningful consultation will remain a superficial method of consulting without accommodating." This argument applies equally to collaboration, co-management, and other "integrative" projects. La Salle (2010, 2013), for example, identifies discrepancies between how collaboration is marketed and how it is actually practised. She asks what I consider to be the critical question: "whether the collaborative research model represents a break with the past, or is instead simply the same old practice dressed up in fashionable new language" (La Salle 2010: 401).

Indeed, the scale of the problem is not one that can be solved by individuals with "good intentions" (La Salle 2010). While there may be "a 'feel good' factor for the discipline at large here, it is often more a politics of gesture than serious recognition" (Smith 2004: 200). The reality is that heritage destruction is deeply embedded in the structure of society (Williams 2012), thus research and (social) science itself, and so "these same individuals, when they are employed or are rewarded by elite power structures that are founded upon racist ideologies, sustain these structures" (Dabulkis-Hunter 2002: 77). As Smith (2004: 198) has asserted, "[w]ithout an active understanding of power/knowledge relations, any attempts to incorporate Indigenous or other non-archaeological knowledges and aspirations into archaeological practice will simply end in appropriation by archaeologists, no matter how unintended that may be." This is what Taiaiake Alfred (2009: 99) describes as co-optation, a process that hinges on appropriating dissonant or "alternative" views into mainstream theories and practices. This process is a replacement for direct coercion.

9 Lack of scientific research is not a limiting factor in resolving the maritime heritage crisis; indeed, because of its values, science is itself a limiting factor because it promotes the ideology of growth, development, and progress.

The mainstream response to the maritime heritage crisis has been a near unanimous call for more research. This presumes there is a need for more data. Yet, government agencies and local communities already have official lists, many of them digitized for quick mapping, that include key information needed to make decisions about at-risk heritage sites. Indeed, for coastal archaeologists and Indigenous communities, virtually all CRM is about dealing with these very issues and places. The problem, then, is not insufficient data. In this sense, the call for more research itself is a call for growth, development, and progress—notions which define the scientific process. One need only add up all the external costs of a large-scale archaeological field research project to see how this unfolds in the real world.

Additionally, since the 1970s the reactions of archaeology to the commercialization process were "not to challenge the commercial environment for archaeology, but to reinforce its foundations by shadowing business principles" (Zorzin 2011: 131). Here, the interest is not in protecting or conserving sites but in profiting from them. In this sense, archaeologists are being paid to oversee heritage destruction. Further, archaeologists working in service of developers are inherently working in service of development. In this world, it may be in one's interest to not find sites, or not value them fairly. This process is mirrored in situations whereby scientists working for government agencies, or interacting with them in other capacities, "are directed to lie, to reinterpret findings in ways that support a policy position that has already been decided upon, or not disclose findings that might be politically inconvenient" (Schrecker 2001: 40). Attempts to incorporate values outside of scientific archaeology, such as the views of Indigenous communities, have so far failed to produce any meaningful change in the system (Markey 2001).

10 "Resources" are defined within a capitalist economy regulated by the state that promotes and is supported by a broader "culture of over-consumption."

Resource management is the prevailing Western environmental paradigm. In Canada, the "primary idea" driving recent resource use, thus policy discourse, has been that of "resource management as the allocation of public resources to private industry" (Hessing and Howlett 1999: 12). CRM, however, does not fall so neatly into this categorization. On one hand, archaeology is recognized as being fundamentally an extractive process, wherein archaeological sites are mined for archaeological resources (artifacts) to produce scientific data and artifact hordes. Once removed from the ground, both archaeologist and museologist translate the resource into economic profit and social power.

On the other hand, rather than being a resource that is extracted and manipulated, cultural heritage can be seen as falling into the category of "resources" that impede economic development. In this latter sense, cultural heritage is like the spotted owl or eelgrass: spotted owl impede timber extraction and eelgrass impedes port construction, and thus both impede development, thus "productivity" (i.e. profit-maximization). In the end, cultural heritage, owl, and eelgrass alike are "managed" in a nearly identical manner as wood and gravel. Through resource management, so-called "cultural" resources are objectified, commodified, then "negotiated." Resources are located within an economic framework based on a market economy with state regulation (Hessing and Howlett 1999: 47) that facilitates a "culture of over-consumption" (Bodley 2008a: 101–20).

11. Resource management is a modern hegemonic project working as designed.

Points 1–10 lead me to conclude that archaeology/CRM is performing as it was designed. In contrast to the goals set out in the HCA of protecting and conserving heritage sites, my research illustrates that the practice of archaeology/CRM, defined by its outcomes, is primarily concerned with clearing Indigenous heritage from the landscape to make way for capitalist development. Archaeology/CRM is the process by which heritage is converted, via science and policy, from a cultural idea to a natural commodity—a resource in need of management. This gives rise to the cultural resource manager, or consulting archaeologist, whose task is to extract cultural heritage for profit.

The provincial government's official position is "not to prohibit or inhibit land use and development, but rather to assist the development industry, the province, regional authorities, and municipalities in making decisions leading to rational land use and development" (BC 2012). From a political economy perspective, such "collaboration with industry" represents the co-optation of archaeology—and its (presumed) conservationist ideals—by profit-motivated developers. The result has been CRM where, superficially, the values of archaeology persist yet, in reality, development not only goes unchallenged, it is integral to the success of the CRM industry. Indeed, for some, this is an optimistic view wherein development-led site destruction "could be turned form a bane to a boon" for the practice (Borden 1950; compare with Doyle 2013; O'Connor 2014). Not only does archaeology/CRM not challenge development, it facilitates and promotes it. In this sense, the institution may be considered to be working perfectly.

6 Looking Forward, Looking Back

> Unalloyed and simplistic optimism about the future is really just denial in another guise.
>
> (Homer-Dixon 2007: 214)

> The revelations are painful, but they must be faced if the present is to be understood, and the future made more tolerable.
>
> (Duberman 1964: 5)

Contrary to the laws of physics, there exists—in the same time and place—two archaeologies. The first is a democratic force, wherein politically neutral archaeologists, working in alliance with Indigenous peoples and other "stakeholders," carry out their practice's prime directive: use science to "preserve the past." In this reality, liberal-minded, eco-friendly university professors and their students work tirelessly and with little reward to save heritage from the ravages of time. Being the dominant or mainstream view, this is most likely the practice you learned about in your Introduction to Archaeology class, or from mass media.

The second archaeology is a tyrannical force, wherein politically motivated archaeologists, working as part of the neoliberal state, control how mass society perceives and interacts with contemporary heritage landscapes, and thus contemporary peoples. In this reality, mostly white middle-class conservative scientists are paid to permit heritage destruction with the goal of clearing heritage—typically that of Indigenous peoples—from the landscape in advance of development. As the minority or marginal view, this is the practice you likely never learned about it in your Introduction to Archaeology class, or from mass media.

It turns out that which view is right—archaeology as democracy, archaeology as tyranny—depends upon your standpoint (Hicks and McAtackney 2007; Kahan et al. 2011). If you are a professional archaeologist working for a university or cultural resource management (CRM) firm, chances are you subscribe to the former. If you are an Indigenous person, chances are you subscribe to the latter.

Looking Back

> And that claim is by the right of our manifest destiny to overspread and to possess the whole of the continent which Providence has given us for the development of the great experiment.
>
> (O'Sullivan 1845: 6)

> It is impossible not to look forward to distant times when our rapid multiplication will expand itself beyond those limits, and cover the whole northern, if not the southern continent.
>
> (Jefferson 1801: n.p.)

I set out in *Maritime Heritage in Crisis* to chart the contours of Indigenous heritage landscape destruction. Like global ecological breakdown, the maritime heritage crisis is the product of modernity, an historical process set in motion many generations ago and reproduced daily by individuals in the present. As indicated in the comments above from the first half of the 19th century, the heritage crisis has been long in the making (Forbes 2008; Williams 2012), with major benchmarks at 1492, 1750, and 1950. In light of the onset of accelerated climate change and the passing of the millennium, I add the year 2000 to that list.

Late modernity is the socioeconomic context under which the maritime heritage crisis has come into being. I have focused in particular here on the late modern period, 1950–present, the age associated with the Great Acceleration and the Management Era. The Great Acceleration is the time when many human activities (e.g. commercial fishing, greenhouse gas emissions, mass media) reached take-off points, sharply accelerating towards the present/future. The period is described as "the most profound transformation of the human relationship with the natural world in the history of humankind" (International Geosphere–Biosphere Programme 2015). It is no coincidence that the Great Acceleration overlaps temporally with Hessing and Howlett's (1999: 50) Management Era, characterized by an "overall increase in the role and powers of the state in regulatory processes, and a corresponding increase in the complexity of legislation and administrative structures."

I have explored in this book the twin impacts of population growth (coastal sprawl) and climate change (sea level rise) on Indigenous maritime heritage landscapes. The logic behind the dual focus was that if one's goal is to understand the impacts of sea level rise on heritage, then one must minimally grapple with development. This is because (*i*) development is the cause of sea level rise and (*ii*) responses to inundation are largely dictated by the prevailing social environment (Barnett and Casper 2001), which includes the existing built environment (i.e. structure and infrastructure).

In Chapter Two, I discussed the correlated problems that are sea level rise and amenity migration, concluding that both collectively constitute a maritime heritage crisis for coastal Indigenous peoples. In Chapter Three, CRM was identified as the mainstream response to the crisis, and shown to be a very recent (post-1950) invention of the state (Hessing and Howlett 1999; Smith 2004; Spurling 1986; West 1995). In British Columbia, the rise of archaeology/CRM corresponds with the rise of British Columbia itself, defined in terms of the province's transition into a modern resource economy. In this project, Indigenous heritage landscapes often constitute a barrier or added cost (Bakan 2004) to doing business that, under the neoliberal regime (Bendix et al. 2012), must be removed.

In Chapter Four, I presented my case study of archaeology/CRM as it has played out historically within the traditional territory of the *shíshálh* Nation. My goal was to quantify and qualify how effective CRM has been in protecting *shíshálh* heritage. Operating from the position that the past can be used as a proxy for the future, I considered this to be a measure of how effective CRM will be in "managing" the two-pronged maritime heritage crisis. This holistic landscape survey was conducted in light of *shíshálh* First Nation heritage stewardship principles, and my analyses revealed the severity of the crisis on the *shíshálh* Coast, summarized in seven key findings:

1 Residential development has replaced logging as the primary cause of site disturbance on the *shíshálh* Coast.
2 At least 75 percent of all known sites examined have some degree of disturbance.
3 90 percent of known sites are located at the contemporary shoreline.
4 For scientific archaeologists, a "disturbed" site is effectively considered destroyed.
5 For the *shíshálh* Nation, all sites are "significant" irrespective of any measure of site destruction or disturbance.
6 While the *shíshálh* have tried to be proactive in the stewardship of their heritage, their ability to affect meaningful change is severely limited by the state and its supporters.
7 If the goal of (state-controlled) archaeology/CRM management is to protect sites, then the *shíshálh* Coast study demonstrates that it is an ineffective strategy.

Critically, the primary cause of *shíshálh* Nation heritage destruction is "external" (i.e. it is extra-local). For so long as this remains the case, "solving" the heritage crisis through local strategies will invariably be unsuccessful.

In Chapter Five, I sought to understand this dynamic of external pressures in order to better discern the nature of the problem, and thus obstacles to its resolution. While my focus on resourcism led me straight to political economy (capitalism) and the Western ideology of progress, my focus on management led directly to power. From this vantage, the maritime heritage crisis is fundamentally a social concern, not a scientific one. Toward that end, I adopted a critical approach to heritage (Smith 2004) which focused on the ideas behind the crisis rather than on its material components.

I turned to John Bodley's power-elite hypothesis, which states that as communities grow in size, social power becomes increasingly concentrated. This dynamic constitutes a feedback loop whereby an ideology of growth, development, and progress dominates. In the case of the *shíshálh* Coast, this ideology and its on-the-ground consequences in the form of development (sprawl) are external to that community. Instead, they derive from the power-elite core, including the economic center of Vancouver and the managerial center that is Victoria.

In this way, control over heritage is maintained by the elite and the state, which operate from the core to manage the periphery—part of what social geographers refer to as the "consumption of rural space" (Cater and Jones 1993: 219); this is the colonial system par excellence.

The outcome of external control is to decouple people from land and history, and disassociate "natural" or environmental heritage from "cultural" heritage. Made possible through the bureaucratic management system, this process reduces community health and wellbeing. It also facilitates the commodification process whereby cultural heritage is reframed as archaeological resources. A form of disaster capitalism (Hutchings 2016; Hutchings and La Salle 2015a, 2015b), archaeology/CRM's main function is to clear Indigenous heritage from the landscape, thereby expropriating land and resources from Indigenous control and making way for economic development. This idea of "clearing" the landscape of Indigenous heritage brings to mind a number of related concepts that fall under the larger umbrella of memory (Benton 2010), specifically collective amnesia (Mapes 2009) and heritage landscape erasure (Arnold 2014; Dearborn and Stallmeyer 2010; Smith 2008). These processes are linked not only to capitalism but to the larger projects of colonialism, imperialism, and modernity.

Working as designed, archaeology is a billion-dollar industry in North America today, with the vast majority of money derived from developers (American Cultural Resources Association [ACRA] 2013). This economic structure guarantees archaeology will work in service of the neoliberal state, the outcome of which is hegemony and the disintegration of local control over heritage. This is important because archaeology as slow violence is directly implicated in the destruction of Indigenous heritage landscapes, with potentially devastating consequences for Indigenous communities whose connections to place are fundamental to the preservation of cultural memory and identity.

Globally, Indigenous peoples are fighting to maintain these connections and are insisting upon control over their resources and places that matter (Council of the Red Nation 2015; Hall and Fenelon 2009; Ridgeway and Jacques 2014; Ross et al. 2011). However, my reading of the *shíshálh* Coast study results is that the *shíshálh* Nation has very little control over their heritage, despite being self-governing and despite recent gains in the arena of cultural heritage stewardship (Nicholas 2006; Yellowhorn 2012). A central issue is thus sovereignty (Asch 2009; West 1995).

The maritime heritage crisis is a philosophical and moral problem, unprecedented in human history in both scope and scale. Yet, Canada, British Columbia, and regional and local governments therein have spent virtually no time or money addressing it. This is not to suggest more money can resolve any of these issues; rather, it indicates to me that the subject of heritage conservation is ultimately of little concern to non-Indigenous governments. To date, for example, there is no study for either British Columbia or Washington State that recognizes the near-complete destruction of maritime heritage landscapes over the next century, as described here for the *shíshálh* Coast, despite the fact that it has been known about for decades (Easton 1990). I find it particularly disconcerting that North American academic archaeologists have largely ignored the crisis. The apparent lack of concern lies in stark contrast to Europe, in particular England (e.g. Council for British Archaeology [CBA] 2007; English Heritage 1999, 2002, 2003, 2006a, 2006b, 2007, 2008, 2010; Fulford et al. 1997).

From my standpoint, state support for heritage conservation in England is due to the fact that virtually all heritage there is conceived of as national heritage, thus making all heritage "English" heritage. Conversely, in British Columbia, heritage post-dating 1846 is viewed as "British Columbian" heritage (i.e. settler heritage). In this construct, "prehistoric" sites (i.e. Indigenous heritage) are primarily conceived of as resources. Despite this difference in state and, therefore, economic support for heritage conservation, there is little evidence to suggest the situation in England is any different than in North America—specifically, the everyday practice of clearing sites for development is ubiquitous.

So, what does the maritime heritage crisis say about archaeology, CRM, and global ecological breakdown? First, archaeology/CRM is not designed to conserve or protect cultural heritage landscapes. I agree with King (2009) that resource management is by and large "a sham." Archaeology continues to whitewash heritage destruction, locally and globally, and few academic archaeologists show interest in talking openly and honestly about CRM. This means that the dominant voices in archaeology are largely uncritical about the practice. By not speaking loudly and clearly, and making CRM the central focus of discourse, its practitioners are complicit in the destruction of Indigenous heritage.

Upon recognizing the crisis that maritime heritage faces, a common response is to reframe the discussion in terms of possible solutions—how do we deal with or manage this crisis? What changes can be made that will fix the problem? Where do we go from here? Yet to have a conversation about "solutions," one must first agree on what exactly the problem is. For archaeologists, the problem may be viewed as the loss of heritage sites without archaeological investigation first being undertaken. For Indigenous communities, the problem may be seen as the loss of control over their heritage, defined both as tangible sites and intangible landscapes. For non-Indigenous governments, the problem may be considered to be technical and bureaucratic—ensuring the "proper" people are involved, the correct forms filled out, the relevant laws and by-laws followed. For developers, and thus for the resource managers they employ, the problem is invariably financial—ensuring the most return for the least investment.

In my view, the problem is the hegemonic power-elite dynamic, which concentrates economic power and fuels the capitalist ideology of growth, development, and progress, now with planetary-scale consequences. There is no quick technical fix to any one of the problems enmeshed in this complex web, for any change tugs on the threads of all in unpredictable and undesirable ways. To be clear, it is not that there are no solutions, but rather these solutions are not in the interests of the power-elite. Thus, under the current economic circumstances (see Box 6.1), the global ecological crisis—of which the maritime heritage crisis is a part—is intractable.

Box 6.1 Global ecological breakdown in perspective: Thinking critically, part II

The destruction of the past is perhaps the greatest of all crimes. ... We must put an end to the terrible uprootedness which European colonial methods always produce, even under their least cruel aspects.

Simone Weil, 1952: 48

Capital's endless pursuit of new outlets for class-based accumulation requires for its continuation the destruction of both pre-existing natural conditions and previous social relations. Class exploitation, imperialism, war, and ecological devastation are not mere unrelated accidents of history but interrelated, intrinsic features of capitalist development.

John Bellamy Foster, 2009: 40–1

There is nothing natural about this behavior. People are not naturally driven to accumulate wealth. There are societies in which such accumulation is discouraged. Human beings do not have an innate drive to accumulate commodities; again, there are plenty of societies in which such accumulation is discouraged. People are not driven to work; in fact, contrary to popular notions, members of capitalist culture work far more than, say, people who live by gathering and hunting.

Richard Robbins, 2005: 13

The *ecological rift* is the rift between humanity and nature. The world is really one indivisible whole.

John Bellamy Foster, Brett Clark, and Richard York, 2010: 7

We need to recognize that we as human beings are a part of nature, not apart from nature. ... An ecological revolution means breaking with the vicious cycle of the exploitation of both people and nature.

Fred Magdoff and John Bellamy Foster, 2011: 144

As a psychopathic creature, the corporation can neither recognize nor act upon moral reasons to refrain from harming others. Nothing it its legal makeup limits what it can do to others in pursuit of its selfish ends, and it is compelled to cause harm when the benefits of doing so outweigh the costs. Only pragmatic concern for its own interests and the laws of the land constrain the corporation's predatory instincts, and often that is not enough to stop it from destroying lives, damaging communities, and endangering the planet as a whole.

Joel Bakan, 2004: 60

Industrial capitalism can never be sustainable. It has always destroyed the land upon which it depends for raw materials, and it always will. Until there is no land (or water, or air) for it to exploit. Or until, and this is obviously the far better option, there is no industrial capitalism.

Derrick Jensen and Aric McBay, 2009: 201

Ironically, most analyses of the environmental problem today are concerned less with saving the planet or life or humanity than saving capitalism—the system at the root of our environmental problems.

John Bellamy Foster, Brett Clark, and Richard York, 2010: 7

I think a question that we are not asking ourselves is: isn't humanity committing suicide with this indiscriminate and tyrannical use of nature? Safeguard creation because, if we destroy it, it will destroy us. Never forget this.

Pope Francis, 2014: n.p. (in Vidal 2015)

I wish to repeat that the ecological crisis is a moral issue.

Pope John Paul II, 1990: n.p.

Looking Forward

> The future is at once both very bright for the employment of maritime archaeologists but very gloomy for maritime cultural heritage.
>
> (Flatman 2009: 6)

> We need to think about moving populations and we need to make new communities.
>
> (Ellis 2015: n.p.)

There is strong evidence to suggest that the maritime heritage crisis is irresolvable, at least in the current state (Bakan 2004; Club of Rome 1970; Council of the Red Nation 2015; Fiala 2010; Moghaddam 1997, 2010). Based on my retheorization of archaeology/CRM in this book, and following Flatman above, I observe an inverse relationship between the number of professional archaeologists and the number of intact Indigenous heritage landscapes, as shown in Figure 6.1. The paradox is unlikely to change anytime soon "if analysis and actions assume capitalism. Conservation is a lost cause as long as it is conceived of as a specialty activity operating within the structures and processes of industrial capitalism" (Rogers 1998: ix). This is because heritage conservation is inseparable from its social contexts. Minimally, that context includes modernity, capitalism, and the state (Hutchings and La Salle 2013, 2015a).

Problematically, mainstream responses to heritage problems "strategize with variables that are part of these problems without necessarily engaging in a critical analysis of those variables" (Rogers 1998: xi):

> The conception of sustainability as "integration" of conservation and development requires challenge on the part of conservation, rather than integration. Historical forces which are marginalizing human communities and destroying natural communities are aggressively pursuing an intensifying agenda of exploitation and domination. Marginalized groups do not find themselves in benign situations, by definition. Therefore, any call for conservation in this context is a call that requires resistance and challenge to current realities, rather than accommodation and integration.

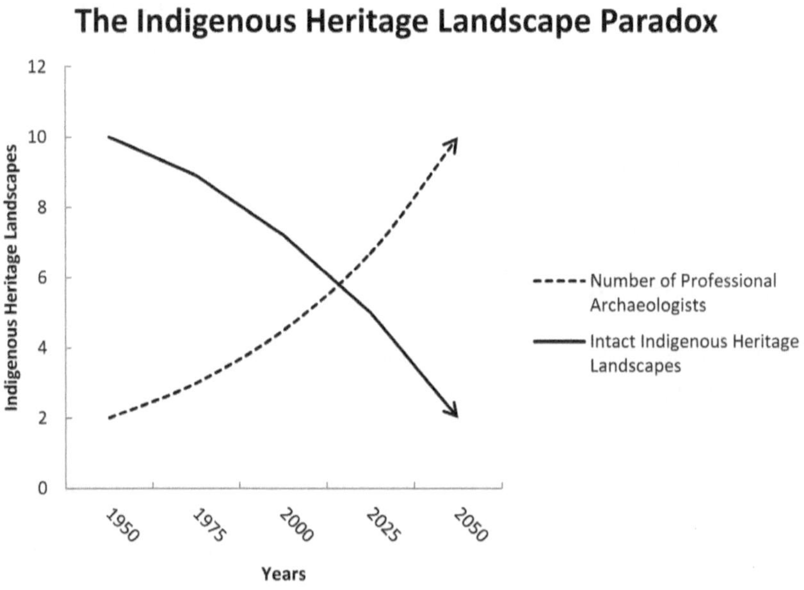

Figure 6.1 There exists a strong relationship between the number of professional archaeologists, both academic and CRM, and the rate and scale of Indigenous heritage landscape destruction. I theorize this relationship here as the Indigenous Heritage Landscape Destruction Paradox.

I posit that "accommodation" and "integration" in natural heritage discourse (Armitage et al. 2007; Rogers 1994, 1995, 1998) map directly onto "consultation" and "collaboration" in cultural heritage discourse (Atalay et al. 2014; Gnecco 2012; Hutchings 2014; King 2009; La Salle 2010, 2014b; La Salle and Hutchings 2015, 2016).

Critical analysis invariably leads to the same conclusion. King, for example, calls the resource management process "a sham." La Salle calls the consultation process a "trade bead." Gnecco (2012: 398) points out that, for most archaeologists:

> "collaboration" is more a way of alleviating their guilt (and getting on with their work) than a way of embarking on the path of different practices; more of the "political correction" that reaches out to marginalized peoples, frequently with an arrogant naiveté built upon selected criteria of authenticity and purity. Many archaeologists are content with offering to local communities cultural crumbs (a local museum, a video, a school booklet) while preserving their control over critical issues (such as research design, curation of findings, production and distribution of archaeological narratives).

As La Salle and I point out in Archaeology as Disaster Capitalism (Hutchings and La Salle 2015a, 2015b), archaeologists as a collective have always been "mercenaries," be it for the state or the corporation.

Given my focus in this book on mainstream responses, it is appropriate to consider different reactions to ecological breakdown. Sociologist Anthony Giddens defines four different "adaptive reactions" to the crisis of modernity (1990: 135–7). The purpose of considering these different responses is to place archaeology/CRM in philosophical context. As illustrated in Figure 6.2, the four responses are: cynical pessimism, pragmatic acceptance, sustained optimism, and radical engagement.

Cynical pessimism "presumes a direct involvement with the anxieties provoked by high consequence dangers" (Giddens 1990: 136):

> Cynicism is not indifference. Nor is it necessarily doom-laden, although it is hardly compatible with blunt optimism. Cynicism is a mode of dampening the emotional impact of anxieties through either a humorous or a world-weary response to them. It lends itself to parody.

In archaeology, cynical pessimists see structural problems, such as capitalism, as unfixable and do not see their work as any consolation. This response, akin to hopelessness, precludes meaningful change.

Pragmatic acceptance involves recognizing the structural problems and concentrating on "surviving":

Pragmatic acceptance	Sustained optimism
Cynical pessimism	Radical engagement

Figure 6.2 Four responses to modernity, thus global ecological breakdown and the maritime heritage crisis.

Source: Data from Giddens (1990: 135–7).

> What is at issue here is not so much a withdrawal from the outside world as a pragmatic participation which maintains a focus on day-to-day problems and tasks. [T]he belief that much that goes on in the modern world is outside anyone's control, so that temporary gains are all that can be planned or hoped for.
>
> (Giddens 1990: 136)

Pragmatic acceptance is "not without psychological costs," as it implies a "numbness frequently reflecting deep underlying anxieties, which in some individuals repeatedly surface at a conscious level." In archaeology/CRM, this is the approach taken by practitioners who see capitalism as outside of their control and so concentrate on just doing "good" archaeology (Hutchings and La Salle 2015c).

Sustained optimism is essentially "the persistence of the attitudes of the Enlightenment"—it is a "continued faith in spite of whatever dangers threaten at the current time."

> This is the outlook of those experts … who have criticised "doomsday" ecological scenarios in favour of the view that social and technological solutions can be found for the major global problems. ... it is a perspective

> which continues to hold great resonance and emotional appeal, based as it is upon a conviction that unfettered rational thought and particularly science offer sources of long-term security that no other orientations can match.
>
> (Giddens 1990: 136)

This is the "SIT and wait" (instrumentalist) approach discussed in Chapter One. Archaeologists taking this position feel structural problems can be fixed through policy changes, science, and new technologies.

Finally, radical engagement refers to "an attitude of practical contestation towards perceived sources of danger":

> Those taking a stance of radical engagement hold that, although we are beset by major problems, we can and should mobilise either to reduce their impact or to transcend them. This is an optimistic outlook. ... Its prime vehicle is the social movement.
>
> (Giddens 1990: 137):

Based on the *shíshálh* Coast study and my understanding of the maritime heritage crisis, radical engagement is clearly what is needed. Yet, I question the possibility of transcending these issues. In addition to my experiences with the Club of Rome's Continuous Critical Problems and the Indigenous Heritage Landscape Paradox, this lack of faith is due in large part to the "issue-attention cycle," shown in Figure 6.3.

The five step *issue-attention cycle* represents people's interest in heritage issues. The cycle begins with a great fanfare around a new idea, giving way to recognition of the full social and economic costs associated with attending to that problem. This is followed by the gradual but inevitable decline of interest, ending in the state Anthony Downs (1972) calls "prolonged limbo." There, problems sit idle and unattended for long periods of time until an individual or group revisits the issue (pre-problem stage), once again to great fanfare. This seems to be the pattern for archaeology in western Canada, where many of the issues addressed here were spelled out decades ago (e.g. Spurling 1986; West 1995). The question, then, becomes why such stories are so quickly and routinely "forgotten." Such *collective amnesia* (Mapes 2009) continues because such narratives constitute for archaeologists an inconvenient truth.

For me, the question of intractability comes down to just one essential issue: truth-telling. Truth-telling is important because it fundamentally influences our baselines, or starting points, thus also our goals. Rationalizing away uncomfortable truths should not and cannot be the norm. On that note, and thinking of the future, I take up this challenge from Ian Angus (2013: 28):

> the first step is to tell the truth—about the danger we face, about its causes, and about the measures that must be taken to turn back the threat. In a time of universal deceit, telling the truth is a revolutionary act.

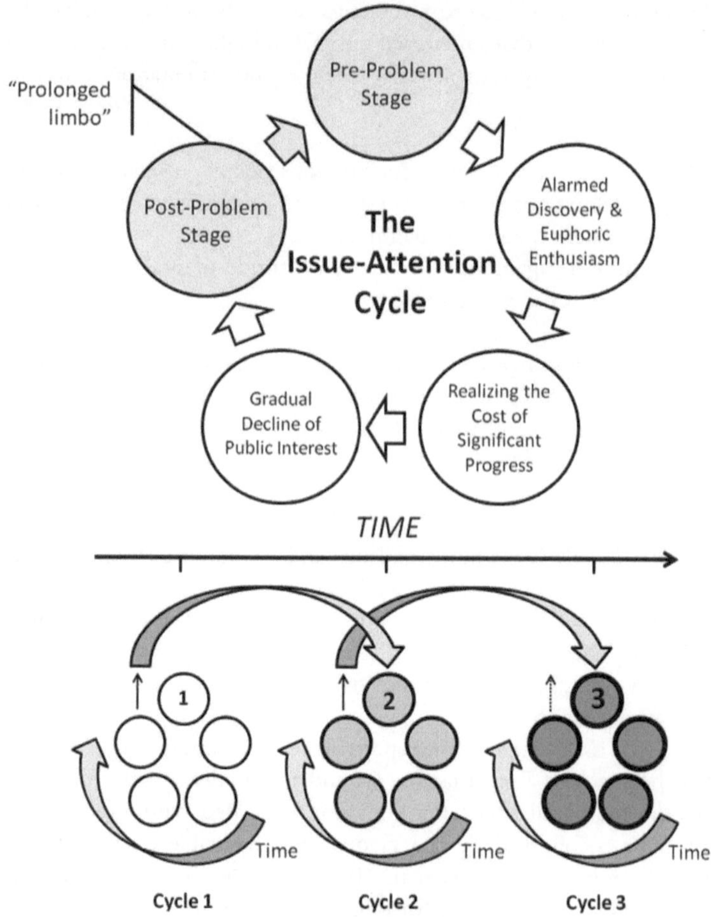

Figure 6.3 The issue-attention cycle in theory (top) and reality (bottom). As illustrated, serious socioenvironmental problems such as the maritime heritage crisis spend most of their time in "prolonged limbo" (grayed stages at top), where they are simply ignored. While the issue gets revisited over time, the outcome remains the same, "decline of interest" and "prolonged limbo." Intractable problems, once revisited, are rarely the same as the previous time they were addressed, as suggested in the top diagram. Rather, as shown at bottom, problems morph, often getting more concatenated and thus intractable through time. This view reflects the increasing cost of "doing nothing."

Source: Adapted from Downs (1972: 39–41) and Mitchell and Sewell (1981: 13, Figure 1.1). Reprinted with permission of Oxford University Press Canada.

Freeing the Canary

If you care to look, you can see the canary alerting us.

(Guinier and Torres 2002: 12)

> Native peoples experience extreme and regular forms of violence because the whole system relies on our death. Capitalism–colonialism means death for Native peoples. For Native peoples to live, capitalism and colonialism must die.
>
> (Council of the Red Nation 2015: 7)

According to the miner's canary metaphor, as initially described in 1953 by Felix Cohen, Indigenous communities play much the same role in contemporary North American society that Jewish peoples played in Nazi Germany. Like the miner's canary, Cohen asserted, Indigenous peoples mark the movements "from fresh air to poison gas in our political atmosphere"—treatment of Indigenous peoples reflects the "rise and fall in our democratic faith" (1953: 390).

I consider the state of Indigenous maritime cultural landscapes to be a proxy for not just the current global sociopolitical "atmosphere," but also the ecological crisis of which we all are now a part. Like the canary that warns the miner of a toxic atmosphere, issues of marginalized heritage—specifically Indigenous heritage—point to underlying, structural problems in Western society that ultimately affect everyone. In this way, heritage is very much like race:

> The canary's more fragile respiratory system would cause it to collapse from noxious gases long before humans were affected, thus alerting the miners to danger. The canary's distress signaled that it was time to get out of the mine. … Those who are racially marginalized are like the miner's canary: their distress is the first sign of a danger that threatens us all. It is easy enough to think that when we sacrifice this canary, the only harm is to the communities of color. Yet others ignore problems that converge around racial minorities at their own peril, for these problems are symptoms warning us that we are all at risk.
>
> (Guinier and Torres 2002: 11)

For Lani Guinier and Gerald Torres, the miner's canary metaphor "captures the association between those who are left out and social justice deficiencies in the larger community" (2002: 11–12). In this sense, the canary is diagnostic, "signaling the need for more systemic critique," that is, more social critique (2002: 12). Specifically:

> The miner's canary metaphor helps us understand why and how race continues to be salient. Racialized communities signal problems with the ways we have structured power and privilege. These pathologies are not located in the canary. Indeed, we reject the … approach that locates complex social and political problems of the mines by outfitting the canary with a tiny gas mask to withstand the toxic atmosphere. … Political race tells us that we need to change the air in the mines. If you care to look, you can see the canary alerting us.
>
> (Guinier and Torres 2002: 12)

Late modern conceptions of "heritage," "environment," and "stewardship" are tied up with—and thus inseparable from—Western conceptions of race, especially in colonial nations like Australia, New Zealand, Canada, and the United States (Armitage 1995). This connection is best exemplified in the idea of institutionalized environmental racism (Gosine and Teelucksingh 2008), which is any institutional "policy, practice or directive that differentially affects or disadvantages (whether intended or unintended) individuals, groups, or communities based on race or colour" (Bullard 1996: 497).

Western land heritage policies are designed to drive a wedge between Indigenous peoples and their "resources" (Bodley 2008b: 95–116, 117–40; see also Council of the Red Nation 2015; Gosine and Teelucksingh 2008: 33–62). As Jerry Mander (1992: 263) puts it, "In the later stages of an epic worldwide struggle, the forces of Western economic development are assaulting the remaining native peoples of the planet, whose presence obstructs their progress." State-sanctioned heritage destruction is part of this larger assault. While most archaeologists seem to be largely oblivious to this reality, it is one Indigenous peoples have struggled with for centuries:

> Native people are under constant assault by a capitalist–colonial logic that seeks the erasure of noncapitalist ways of life. Colonial economies interrupt cooperation and association and force people instead into hierarchical relations with agents of colonial authority who function as a permanent occupying force on Native lands. These agents are in place to enforce and discipline Native peoples to ensure that we comply with capitalist–colonial logics. ... These are all violent forms of social control and invasion that extract life from Natives and other oppressed peoples in order to increase profit margins and consolidate power in the hands of wealthy nation-states like the USA. The whole system depends on violence to facilitate the accumulation of wealth and power and to suppress other, non-capitalist ways of life that might challenge dominant modes of power.
>
> (Council of the Red Nation 2015: 6–7)

These dynamics are prominent in British Columbian archaeology, where most archaeologists are white and serve the state; meanwhile, archaeology almost exclusively targets Indigenous people. As such, it may be appropriate to consider archaeology/CRM a form of heritage racism, which disenfranchises Indigenous peoples from their heritage, resulting in solastalgia.

Given the systemic nature of the crisis, what is to be done? A useful place to start is with Zoltán Grossman and Allen Parker's (2012) book *Asserting Native Resilience: Pacific Rim Indigenous Nations Face the Climate Crisis*. Their collection of papers and stories from the Pacific Northwest Coast is particularly relevant because it focuses on sea level rise and draws on the miner's canary metaphor. In his essay No Longer the "Miner's Canary," Grossman (2012) considers Indigenous nations' responses to climate change and concludes that "the fate of Native peoples provides an early warning to the fate of all humanity":

> On one hand, Indigenous peoples are on the front line of climate change—the first to feel its effects. … On the other hand, Indigenous peoples have certain advantages in responding to climate change, compared to non-Native neighbors or local governments. They want their continuing life-ways (not their death) to provide some direction to the rest of humanity. Perhaps the "miner's canary" formulation could be revised, so the canary escapes the cage, flaps its wings, and shows the hapless miner a safe way out of the toxic mine. Indigenous knowledge and experience has the potential to help the rest of humanity get more out of harm's way.
>
> (Grossman 2012: 175)

For Grossman, it is critical that Indigenous climate change discourse not only warns Indigenous communities of the dangers inherent to ecological breakdown, but also empowers those communities around the "inherent tribal strengths" that are political sovereignty and a sense of community. Political sovereignty is a strength because Indigenous communities have a unique status as nations; with their sovereignty, Indigenous communities "can develop their own models of dealing with climate change and managing nature in a sustainable way" (2012: 176). A sense of community is an asset because, in contrast to much of non-Indigenous society, "Indigenous peoples still have a community." These and other strengths play into a wide range of relationships and strategies related to building climate change resilience (2012: 176–85).

Those strengths also play into Alan Parker's (2012: 189–92) ten recommendations to Indigenous government leadership for building resilience to the climate crisis. Parker prefaces those recommendations with the following statement:

> As one contemplates acceleration of climate change in the next five to ten years, our tribal communities must adapt to these changing conditions at a pace that will stress their social, economic, and cultural fabrics. But, we cannot afford to join [mainstream society] in massive denial. The time to plan and adapt is now.
>
> (Parker 2012: 189)

Parker's ten recommendations are:

1 Gather information on the impacts of climate change in your region and make it available to your tribal community.
2 Secure sources of water.
3 Secure sources of food.
4 Prepare for impacts on plant and animal species.
5 Develop relationships with neighboring governments and communities regarding disaster planning.
6 Consider political alliances to build a renewable energy policy.
7 Consider strategies to unite Tribes and First Nations around the protection needed to defend treaty rights.

8 Consider active involvement as sovereign Indigenous governments in global climate change negotiations.
9 Get youth involved in cultural education and defending the future of their nation.
10 Work with other Indigenous nations across imposed colonial boundaries on the basis of being natural regions (such as the Pacific Rim, Northwest Coast, and Salish Sea).

The maritime heritage crisis, then, is alerting all to the global consequences of unfettered growth, development, and progress. In this dynamic, Indigenous peoples consistently bear the brunt of environmental injustice, despite having very little to do with causing it. Both because of the intrinsic link between heritage and culture, and between culture and land, coastal change is a "potential culture killer" (Parker and Grossman 2012: 14). Like other coastal groups, the *shíshálh* run the risk of becoming environmental refugees on their own coast, especially if steps are not taken to secure adequate lands and resources. A major concern in the future will be the legal implications of sea level rise, particularly for Indigenous communities; in British Columbia, it is not only heritage sites but contemporary villages and Indian Reserves that will be inundated.

Until the *shíshálh* Nation has sovereignty over its territory, the protection of their places that matter is unfeasible. To that end, I endorse John Bodley's (2013) "small nation solution" as a useful point of departure for talking and thinking about how scale relates to resource management. A critical element of the small nation thesis is downscaling, bolstered by the redistribution of social and economic power. Minimally, downscaling involves nations finding their "optimum size" (2013: vii). Like before, Bodley's premise is that virtually all modern socioenvironmental problems are problems of scale, this because "we are living with systems that we have allowed to grow so large that they cannot be safely managed by even the best intended elites, whether individual investors, business executives, philanthropists, politicians, or the most skilled technicians" (2013: 36). The central problem of large (or industrial) scale is the erosion of democracy, a consequence of the power-elite dynamic, and the concentration of social power. In the context of the present discussion, downscaling involves dismantling national and provincial governments and refocusing on local control over local issues in local landscapes. This involves challenging the state and archaeology/CRM for control over heritage.

Failing such holistic, radical transformation, solastalgia and dislocation will become new baselines. This elevates the need to integrate philosophies of social work (e.g. Albrecht et al. 2007; Alexander 2008; Coates 2003) into theoretical and practical approaches to heritage, which is a fundamental part of human health and wellbeing.

In my final evaluation, the maritime heritage crisis is part of what Raymond Rogers (1994) calls the "crisis of modernity." As such, resolving the crisis requires moving away from the ideologies that underpin late modernity—notably, capitalism and the ideology of unrestrained growth, development, and progress. To do so represents a revolutionary paradigm shift.

Appendix

The Club of Rome's Forty-Nine Critical Continuous Problems (1970: 14–16)

1 Explosive population growth with consequent escalation of social, economic, and other problems.
2 Widespread poverty throughout the world.
3 Increase in the production, destructive capacity, and accessibility of all weapons of war.
4 Uncontrolled urban spread.
5 Generalized and growing malnutrition.
6 Persistence of widespread illiteracy.
7 Expanding mechanization and bureaucratization of almost all human activity.
8 Growing inequalities in the distribution of wealth throughout the world.
9 Insufficient and irrationally organized medical care.
10 Hardening discrimination against minorities.
11 Hardening prejudices against differing cultures.
12 Affluence and its unknown consequences.
13 Anachronistic and irrelevant education.
14 Generalized environmental deterioration.
15 Generalized lack of agreed-on alternatives to present trends.
16 Widespread failure to stimulate man's [sic] creative capacity to confront the future.
17 Continuing deterioration of inner-cities or slums.
18 Growing irrelevance of traditional values and continuing failure to evolve new value systems.
19 Inadequate shelter and transportation.
20 Obsolete and discriminatory income distribution system(s).
21 Accelerating wastage and exhaustion of natural resources.
22 Growing environmental pollution.
23 Generalized alienation of youth.
24 Major disturbances of the world's physical ecology.
25 Generally inadequate and obsolete institutional arrangements.

26 Limited understanding of what is "feasible" in the way of corrective measures.
27 Unbalanced population distribution.
28 Ideological fragmentation and semantic barriers to communication between individuals, groups, and nations.
29 Increasing a-social and anti-social behavior and consequent rise in criminality.
30 Inadequate and obsolete law enforcement and correctional practices.
31 Widespread unemployment and generalized under-employment.
32 Spreading "discontent" throughout most classes of society.
33 Polarization of military power and psychological impacts of the policy of deterrence.
34 Fast obsolescing political structures and processes.
35 Irrational agricultural practices.
36 Irresponsible use of pesticides, chemical additives, insufficiently tested drugs, fertilizers, etc.
37 Growing use of distorted information to influence and manipulate people.
38 Fragmented international monetary system.
39 Growing technological gaps and lags between developed and developing areas.
40 New modes of localized warfare.
41 Inadequate participation of people at large in public decisions.
42 Unimaginative conceptions of world-order and of the rule of law.
43 Irrational distribution of industry supported by policies that will strengthen the current patterns.
44 Growing tendency to be satisfied with technological solutions for every kind of problem.
45 Obsolete system of world trade.
46 Ill-conceived use of international agencies for national or sectoral ends.
47 Insufficient authority of international agencies.
48 Irrational practices in resource investment.
49 Insufficient understanding of Continuous Critical Problems, of their nature, their interactions and of the future consequences both they and current solutions to them are generating.

References

Abrams, Lindsay. 2015a. Visions of a Fallen World: What Our Crowded, Imperiled Planet Really Looks Like. *Salon* 02 March 2015. Electronic document, www.salon.com/2015/03/02/visions_of_a_fallen_world_what_our_crowded_imperiled_planet_really_looks_like/, accessed 16 March 2015.

Abrams, Lindsay. 2015b. When Humans Go Extinct: How Life Will Evolve after We're Gone. *Salon* 14 March 2015. Electronic document, www.salon.com/2015/03/14/when_humans_go_extinct_how_life_will_evolve_after_were_gone/, accessed 16 March 2015.

Acheson, James M. 2006. Institutional Failure in Resource Management. *Annual Review of Anthropology* 35: 117–34.

Acheson, Steve, and Syd Riley. 1979. An Archaeological Resources Inventory of the Northeast Gulf of Georgia Region. In *Annual Report for the Year 1977: Activities of the Provincial Archaeologist Office of British Columbia and Selected Research Reports*, edited by B. O. Simonsen, pp. 39–70. Victoria, BC: Government of British Columbia.

Albrecht, Glenn. 2005. Solastalgia. A New Concept in Human Health and Identity. *Philosophy Activism Nature* 3: 41–55.

Albrecht, Glenn, G. M. Sartore, L. Connor, N. Higginbotham, S. Freeman, B. Kelly, H. Stain, A. Tonna, and G. Pollard. 2007. Solastalgia: The Distress Caused by Environmental Change. *Australasian Psychiatry* 15(1 Supplement): S95–8.

Alexander, Bruce K. 2008. *The Globalisation of Addiction: A Study in Poverty of the Spirit.* New York: Oxford University Press.

Alfred, Taiaiake. 2009. *Peace, Power, Righteousness: An Indigenous Manifesto.* 2nd Edition. New York: Oxford University Press.

Allen, Harry. 2010. The Crisis of 21st Century Archaeological Heritage Management. In *Bridging the Divide: Indigenous Archaeology into the 21st Century*, edited by C. Phillips and H. Allen, pp. 49–60. Walnut Creek, CA: Left Coast Press.

Althusser, Louis. 1971. Ideology and Ideological State Apparatuses (Notes Towards an Investigation). In *Lenin and Philosophy and Other Essays*, edited by L. Althusser, pp. 121–76. New York: Monthly Review Press.

American Cultural Resources Association (ACRA). 2013. *The Cultural Resources Management Industry: Providing Critical Support for Building Our Nation's Infrastructure through Expertise in Historic Preservation.* Washington, DC: American Cultural Resources Association.

Anderson, Benedict. 1983. *Imagined Communities: Reflections on the Origin and Spread of Nationalism.* London and New York: Verso.

Andrews, Thomas D., and Susan Buggey. 2012. Canadian Aboriginal Cultural Landscapes in Praxis. In *Managing Cultural Landscapes*, edited by K. Taylor and J. L. Lennon, pp. 253–71. New York: Routledge.

Angus, Ian. 2013. The Myth of 'Environmental Catastrophism.' *Monthly Review* 65(4): 15–28.

Archaeological Society of British Columbia. 2011. ASBC's Statement on Marpole. *The Midden* 44(2): 1, 4.

Armitage, Andrew. 1995. *Comparing the Policy of Aboriginal Assimilation: Australia, Canada and New Zealand*. Vancouver: UBC Press.

Armitage, Derek, Fikret Berkes, and Nancy Doubleday, eds. 2007. *Adaptive Co-Management: Collaboration, Learning, and Multi-Level Governance*. Vancouver: UBC Press.

Arnold, Bettina. 2014. Erasure of the Past. In *Encyclopedia of Global Archaeology*, edited by C. Smith, pp. 2441–8. New York: Springer.

Asch, Michael. 2009. Concluding Thoughts and Fundamental Questions. In *Protection of First Nations Cultural Heritage: Laws, Policy, and Reform*, edited by C. Bell and R. K. Paterson, pp. 394–411. Vancouver: UBC Press.

Atalay, Sonya, Lee. R. Clauss, Randall H. McGuire, and John R. Welch, eds. 2014. *Transforming Archaeology: Activist Practices and Prospects*. Walnut Creek, CA: Left Coast Press.

Ausenco Sandwell. 2011a. Guidelines for Management of Coastal Flood Hazard Land Use: Climate Change Adaption Guidelines for Sea Dikes and Coastal Flood Hazard Land Use. 27 January 2011. BC Ministry of Environment, Project No. 143111. Victoria: Province of British Columbia.

Ausenco Sandwell. 2011b. Sea Dike Guidelines: Climate Change Adaption Guidelines for Sea Dikes and Coastal Flood Hazard Land Use. 27 January 2011. BC Ministry of Environment, Project No. 143111. Victoria: Province of British Columbia.

Ausenco Sandwell. 2011c. Draft Policy Discussion Paper: Climate Change Adaption Guidelines for Sea Dikes and Coastal Flood Hazard Land Use. 27 January 2011. BC Ministry of Environment, Project No. 143111. Victoria: Province of British Columbia.

Australian National University. 2009. Implications of Climate Change for Australia's World Heritage Properties: A Preliminary Assessment. Report to the Department of Climate Change and the Department of the Environment, Water, Heritage and the Arts by the Fenner School of Environment and Society, the Australian National University. Canberra, Australia: Commonwealth of Australia.

Bakan, Joel. 2004. *The Corporation: The Pathological Pursuit of Profit and Power*. New York: Penguin.

Baran, Paul A., and Paul M. Sweezy. 1966. *Monopoly Capital: An Essay on the American Economic and Social Order*. New York: Monthly Review Press.

Bardsley, Douglas K., and Graeme J. Hugo. 2010. Migration and Climate Change: Examining Thresholds of Change to Guide Effective Adaptation Decision-Making. *Population and Environment* 32(2): 238–62.

Barman, Jean. 2005. *Stanley Park's Secret: The Forgotten Families of Whoi Whoi, Kanaka Ranch and Brockton Point*. Madeira Park, BC: Harbour Publishing.

Barnett, Elizabeth, and Michelle Casper. 2001. A Definition of 'Social Environment.' *American Journal of Public Health* 91(3): 465.

Barry, John. 2007. *Environment and Social Theory*. 2nd Edition. New York: Routledge.

Barth, Roland S. 2013. Culture in Question. In *The Jossey-Bass Reader on Educational Leadership*, edited by M. Grogan, pp. 197–206. San Francisco, CA: Jossey-Bass.

Barthel-Bouchier, Diane. 2013. *Cultural Heritage and the Challenge of Sustainability*. Walnut Creek, CA: Left Coast Press.

Bates, Diane C. 2002. Environmental Refugees? Classifying Human Migrations Caused by Environmental Change. *Population and Environment* 23(5): 465–77.

Battiste, Marie, and James (Sa'ke'j) Youngblood Henderson. 2000. *Protecting Indigenous Knowledge and Heritage: A Global Challenge*. Saskatoon: Purich Publishing.

BC Stats. 2011a. *Overview of the BC and Regional Population Projections, 2011 to 2036*. Victoria: BC Stats.

BC Stats. 2011b. *British Columbia in 2036*. Victoria: BC Stats.

Bendix, Regina F., Aditya Eggert, and Arnika Peselmann, eds. 2012. *Heritage Regimes and the State*. Göttingen: Universitätsverlag Göttingen.

Benton, Tim, ed. 2010. *Understanding Heritage and Memory*. Manchester: Manchester University Press.

Berenfeld, Michelle L. 2008. Climate Change and Cultural Heritage: Local Evidence, Global Responses. *The George Wright Forum* 25(2): 66–82.

Berenfeld, Michelle L. 2015. Planning for Permanent Emergency: "Triage" as a Strategy for Managing Cultural Resources threatened by Climate Change. *The George Wright Forum* 32(1): 5–12.

Berkes, Fikret. 2008. *Sacred Ecology*. 2nd Edition. New York: Routledge.

Berkes, Fikret, Derek Armitage, and Nancy Doubleday. 2007. Synthesis: Adapting, Innovating, Evolving. In *Adaptive Co-Management: Collaboration, Learning, and Multi-Level Governance*, edited by D. Armitage, F. Berkes, and N. Doubleday, pp. 308–27. Vancouver: UBC Press.

Billé, Raphaël. 2008. Integrated Coastal Zone Management: Four Entrenched Illusions. *SAPIENS* 1(2): 1–12.

Birch, Jennifer. 2007. Public Archaeology and the Cultural Resource Management Industry in Ontario. In *Partners to the Past: Proceedings of the 2005 Ontario Archaeological Society Symposium*, edited by J. S. Molnar, pp. 120–7. Ottawa: The Ottawa Chapter of the OAS.

Biro, Andrew, ed. 2011. *Critical Ecologies: The Frankfurt School and Contemporary Environmental Crises*. Toronto, Buffalo, and London: University of Toronto Press.

Blankholm, Hans P. 2009. Long-Term Research and Cultural Resource Management Strategies in Light of Climate Change and Human Impact. *Arctic Anthropology* 46(1–2): 17–24.

Blaser, Mario, Harvey A. Feit, and Glenn McRae. 2004. *In the Way of Development: Indigenous Peoples, Life Projects and Globalization*. London: Zed Books.

Bodley, John H. 1999. Socioeconomic Growth, Culture Scale, and Household Well-Being: A Test of the Power-Elite Hypothesis. *Current Anthropology* 40(5): 595–620.

Bodley, John H. 2001. Growth, Scale, and Power in Washington State. *Human Organization* 60(4): 367–79.

Bodley, John H. 2002a. Anthropology and Global Environmental Change. In *Encyclopedia of Global Environmental Change, Vol. 5: Social and Economic Dimensions of Global Environmental Change*, edited by P. Timmerman, pp. 163–7. Chichester: Wiley.

Bodley, John H. 2002b. Globalization in Historical Perspective. In *Encyclopedia of Global Environmental Change, Vol. 5: Social and Economic Dimensions of Global Environmental Change*, edited by P. Timmerman, pp. 73–85. Chichester: Wiley.

Bodley, John H. 2003. *The Power of Scale: A Global History Approach*. New York: M. E. Sharpe.

Bodley, John H. 2008a [1975]. *Victims of Progress*. 5th Edition. Lanham, MD: AltaMira Press.

Bodley, John H. 2008b. *Anthropology and Contemporary Human Problems*. 5th Edition. Lanham: AltaMira Press.

Bodley, John H. 2013. *The Small Nation Solution: How the World's Smallest Nations Can Solve the World's Biggest Problems*. Lanham, MD: AltaMira Press.

Borden, Charles E. 1950. Letter to Dr. Harry B. Hawthorn, Nov. 10, 1950, Royal British Columbia Museum, Victoria, British Columbia. Copy sent to Wilson Duff, British Columbia Provincial Museum, Victoria, British Columbia. Document on file at UBC Archives, University of British Columbia, Vancouver.

Boucquey, Noëlle, Lisa M. Campbell, Gabriel Cumming, Zoë A. Meletis, Carla Norwood, and Joshua Stoll. 2012. Interpreting Amenities, Envisioning the Future: Common Ground and Conflict in North Carolina's Rural Coastal Communities. *GeoJournal* 77(1): 83–101.

British Columbia (BC). 1996. Heritage Conservation Act [RSBC 1996 Chapter 187]. Electronic document, www.bclaws.ca/civix/document/id/complete/statreg/96187_01, accessed 18 June 2015.

British Columbia (BC). 2006. Alive and Inseparable: British Columbia's Coastal Environment. Victoria: British Columbia Ministry of Environment, Strategic Policy Division.

British Columbia (BC). 2007. Environmental Trends in British Columbia: 2007–Population and Economic Activities. Victoria: British Columbia Ministry of Environment.

British Columbia (BC). 2012. The Archaeological Preservation Process in British Columbia. Victoria: Archaeology Branch, Ministry of Forests, Lands and Natural Resource Operations. Electronic document, www.for.gov.bc.ca/archaeology/preservation_process/index.htm, accessed 18 June 2015.

British Columbia (BC). 2013a. Sea Level Rise Adaptation Primer: A Toolkit to Build Adaptive Capacity on Canada's Southern Coasts (January 2013). Victoria: British Columbia Ministry of Environment.

British Columbia (BC). 2013b. Sea Level Rise Adaptation Primer: A Toolkit to Build Adaptive Capacity on Canada's Southern Coasts (Fall 2013). Victoria: British Columbia Ministry of Environment.

Bruun, Per. 1962. Sea-Level Rise as a Cause of Shoreline Erosion. *Journal of the Waterways and Harbour Division* 88: 117–30.

Bruun, Per. 1983. Review of Conditions for Uses of the Bruun Rule of Erosion. *Coastal Engineering* 7(1): 77–89.

Buggey, Susan. 1999. An Approach to Aboriginal Cultural Landscapes. Historic Sites and Monuments Board Agenda Paper #1999-10. Ottawa: HSMBC.

Buggey, Susan. 2004. An Approach to Aboriginal Cultural Landscapes in Canada. In *Northern Ethnographic Landscapes: Perspectives from Circumpolar Nations*, edited by R. Krupnik, R. Masen, and T. Horton, pp. 17–44. Washington, DC: Arctic Studies Center, National Museum of Natural History, Smithsonian Institute.

Bullard, Robert D. 1996. *Unequal Protection: Environmental Justice and Communities of Color*. San Francisco, CA: Sierra Club.

Caffrey, Maria, and Rebecca Beavers. 2008. Protecting Cultural Resources in Coastal U.S. National Parks from Climate Change. *The George Wright Forum* 25(2): 86–97.

Caffrey, Maria, and Rebecca Beavers. 2013. Planning for the Impact of Sea-Level Rise on U.S. National Parks. *Park Science* 30(1): 6–13.

Campbell, Chief Ernest C. 2012. Open Letter to Premier Christy Clark and Mayor Gregor Robertson, 18 April 2012. Electronic document, www.musqueam.bc.ca/sites/default/files/open_ltr_premier_mayor_marpole_april_18_2012_final_1.pdf, accessed 20 April 2012.

Carlson, Keith Thor, ed. 2001. *A Stó:lō Coast Salish Historical Atlas*. Vancouver: Douglas and McIntyre.

Carson, Rachel. 1962. *Silent Spring*. New York, Houghton Mifflin.

Cassar, May. 2005. *Climate Change and the Historic Environment*. London: Centre for Sustainable Heritage, University College London.

Cassar, May, Christopher Young, Tony Weighell, David Sheppard, Bastian Bomhard, and Pedro Rosabal (in collaboration with the World Heritage Centre and its Advisory Bodies). 2006. Predicting and Managing the Effects of Climate Change on World Heritage. Electronic document, http://whc.unesco.org/uploads/news/documents/news-262-1.doc, accessed 09 May 2013.

Casteel, Richard W., and George I. Quimby. 1975. Introduction: Maritime Anthropology. In *Maritime Adaptations of the Pacific*, edited by R. W. Casteel and G. I. Quimby, pp. 1–4. The Hague: Mouton.

Cater, John, and Trevor Jones. 1993. *Social Geography: An Introduction to Contemporary Issues*. London and New York: Edward Arnold.

Cazenave, Anny. 2014. Anthropogenic Global Warming Threatens World Cultural Heritage. *Environmental Research Letters* 9(5): 051001.

Chapman, Henry P. 2002. Global Warming: The Implications for Sustainable Archaeological Resource Management. *Conservation and Management of Archaeological Sites* 5(4): 241–5.

Charles, Frances G. 2010. One Tribe's Story of Discovery, Conflict and Heartache. *The Midden* 42(2–1): 19–23.

Chen, Jiyu. 2005. Dams, Effects on Coasts. In *Encyclopedia of Coastal Science*, edited by M. Schwartz, pp. 357–9. Dordrecht: Springer.

Clifford, James. 2007. Varieties of Indigenous Experiences: Diasporas, Homelands, Sovereignties. In *Indigenous Experiences Today*, edited by M. de la Cadena and O. Starn, pp. 197–223. Oxford: Berg Publishers.

Club of Rome. 1970. *The Predicament of Mankind: Quest for Structured Response to Growing World-Wide Complexities and Uncertainties*. Rome: The Club of Rome.

Coates, John. 2003. *Ecology and Social Work: Toward a New Paradigm*. Halifax: Fernwood.

Cobb, Miranda. 2010. Meaningful Consultation: Nation to Nation or Domination and Assimilation? MA thesis, Institute of Political Economy, Carleton University, Ottawa, Ontario.

Cohen. Felix S. 1953. The Erosion of Indian Rights, 1950–1953: A Case Study in Bureaucracy. *Yale Law Journal* 62: 348–90.

Cole, Douglas. 1985. *Captured Heritage: The Scramble for Northwest Coast Artifacts*. Vancouver: Douglas and McIntyre.

Collins, Lloyd R. 1953. Archaeological Survey of the Oregon Coast from June 1951–December 1952. Unpublished manuscript on file at the Oregon State Museum of Anthropology. University of Oregon, Eugene, Oregon.

Collins, Randall, and Michael Makowsky. 1998. *The Discovery of Society*. 6th Edition. New York: McGraw Hill.

Coombe, Rosemary J. 2012. Managing Cultural Heritage as Neoliberal Governmentality. In *Heritage Regimes and the State*, edited by R. F. Bendix, A. Eggert, and A. Peselmann, pp. 375–87. Göttingen: Universitätsverlag Göttingen.

Cooper, J. Andrew G., and Orrin H. Pilkey. 2004. Sea-Level Rise and Shoreline Retreat: Time to Abandon the Bruun Rule. *Global and Planetary Change* 43(3–4): 157–71.

Council for British Archaeology (CBA). 2007. *Adapting Archaeology: Foresight for Climate Change in the UK*. York: Council for British Archaeology.

Council of the Red Nation. 2015. Native Liberation Struggles in North America: The Red Nation 10-point Program. *Capitalism Nature Socialism* 26(2): 1–7.

Creel, Liz. 2003. *Ripple Effects: Population and Coastal Regions. Population Reference Bureau, September 2003.* Washington, DC: Population Reference Bureau.

Cronin, J. Keri. 2011. *Manufacturing National Park Nature: Photography, Ecology, and the Wilderness Industry of Jasper.* Vancouver: UBC Press.

Crossett, Kristen M., Thomas J. Culliton, Peter C. Wiley, and Timothy R. Goodspeed. 2004. Population Trends Along the Coastal United States: 1980-2008. Coastal Trends Report Series. Washington, DC: US Department of Commerce, National Oceanic and Atmospheric Administration and National Ocean Service.

Cruikshank, Julie. 2005. *Do Glaciers Listen? Local Knowledge, Colonial Encounters, and Social Imagination.* Vancouver: UBC Press.

Culhane, Dara. 1998. *The Pleasure of the Crown: Anthropology, Law and First Nations.* Vancouver: Talon Books.

Cunliffe, Barry. 2008. *Europe between the Oceans: Themes and Variations (9000 BC–AD 1000).* New Haven, CT: Yale University Press.

Cutting, Robert H., Lawrence B. Cahoon, and Jack C. Hall. 2011. If the Tide is Rising, Who Pays for the Ark? *Coastal Management* 39(3): 282–95.

Dabulkis-Hunter, Susan. 2002. *Outsider Research: How White Writers Explore Native Issues, Knowledge and Experiences.* Palo Alto, CA: Academica Press.

Davis, Richard A., Jr. 2005. Human Impact on Coasts. In *Encyclopedia of Coastal Science*, edited by M. Schwartz, pp. 530–5. Dordrecht: Springer.

Dearborn, Lynne, and John C. Stallmeyer. 2010. *Inconvenient Heritage: Erasure and Global Tourism at Luang Prabang.* Walnut Creek, CA: Left Coast Press.

Deloria, Vine, Jr. 1969. *Custer Died for Your Sins: An Indian Manifesto.* New York: MacMillan.

Deloria, Vine, Jr. 1997. *Red Earth, White Lies: Native Americans and the Myth of Scientific Fact.* New York: Scribner.

Deloria, Vine, Jr. 2012. *Indians of the Pacific Northwest: From the Coming of the White Man to the Present Day.* Golden, CO: Fulcrum.

Dent, Joshua. 2012. Past Tents: Temporal Themes and Patterns of Provincial Archaeological Governance in British Columbia and Ontario. MA thesis, Department of Anthropology, University of Western Ontario, London, Ontario.

Dethier, Megan. 2010. Overview of the Ecology of Puget Sound Beaches. In *Puget Sound Shorelines and the Impacts of Armoring—Proceedings of a State of the Science Workshop, May 2009*, edited by H. Shipman, M. N. Dethier, G. Gelfenbaum, K. L. Fresh, and R. S. Dinicola, pp. 35–41. US Geological Survey Scientific Investigations Report 2010-5254. Reston, VA: US Geological Survey.

Dincauze, Dena F. 2000. *Environmental Archaeology: Principles and Practice.* Cambridge: Cambridge University Press.

Dolan, A. Holly, and Rosemary Ommer. 2008. Climate Change and Community Health: Lessons from Canada's East and West Coasts. *Journal of Rural and Community Development* 3(2): 27–46.

Downs, Anthony. 1972. Up and Down with Ecology—The 'Issue-Attention Cycle.' *Public Interest* 28: 38–50.

Doyle, Alistair. 2013. Pre-Viking Tunic Found by Glacier as Warming Aids Archaeology. Reuters 21 March 2013. Electronic document, http://uk.reuters.com/article/2013/03/21/us-climate-archaeology-idUKBRE92K0V320130321, accessed 18 June 2015.

Duberman, Martin B. 1964. *In White America.* New York: Signet.

Dunnell, Robert C. 1984. The Ethics of Archaeological Significance Decisions. In *Ethics and Values in Archaeology*, edited by E. Green, pp. 61–74. New York: Free Press.

Easton, Norm. 1990. The Accelerated Inundation of Coastal Archaeological Resources in the Pacific Northwest – the Greenhouse Effect and the Development of Underwater Archaeology. Preprint Proceedings of the Circum-Pacific Prehistory Conference, Volume 5, Session X: Future Orientated Pacific Prehistory Research.

Echo-Hawk, Roger, and Walter R. Echo-Hawk. 1994. *Battlefields and Burial Grounds: The Indian Struggle to Protect Ancestral Graves in the United States*. Minneapolis, MN: Lerner Publishing Group.

Ellis, Hugh. 2015. Hull Could Be Wiped Off the Map in 100 Years if Sea Levels Continue to Rise at Current Rate, Warns Expert. *The Independent*, 29 May 2015. Electronic document, www.independent.co.uk/news/uk/home-news/experts-fear-for-hulls-future-as-sea-levels-continue-to-rise-10285219.html, accessed 01 June 2015.

English Heritage. 1999. *A Brief for English Heritage: Rapid Coastal Zone Assessment Surveys*. Swindon: English Heritage.

English Heritage. 2002. *Taking to the Water: English Heritage's Initial Policy for The Management of Maritime Archaeology in England*. Swindon: English Heritage.

English Heritage. 2003. *Coastal Defence and the Historic Environment: English Heritage Guidance*. Swindon: English Heritage.

English Heritage. 2006a. *Ports: The Impact of Development on the Maritime Historic Environment*. Swindon: English Heritage.

English Heritage. 2006b. *Shoreline Management Plan Review and the Historic Environment: English Heritage Guidance*. Swindon: English Heritage.

English Heritage. 2007. *A Brief for English Heritage: Rapid Coastal Zone Assessment Surveys*. Swindon: English Heritage.

English Heritage. 2008. *Climate Change and the Historic Environment*. Swindon: English Heritage.

English Heritage. 2010. *Flooding and Historic Buildings*. Swindon: English Heritage.

Erlandson, Jon M. 2001. The Archaeology of Aquatic Adaptations: Paradigms for a New Millennium. *Journal of Archaeological Research* 9(4): 287–350.

Erlandson, Jon M. 2012. As the World Warms: Rising Seas, Coastal Archaeology, and the Erosion of Maritime History. *Journal of Coastal Conservation* 16(2): 137–42.

Erlandson, Jon M., and Torben C. Rick. 2008. Archaeology, Marine Ecology, and Human Impacts on Marine Environments. In *Human Impacts on Ancient Marine Systems: A Global Perspective*, edited by T. C. Rick and J. M. Erlandson, pp. 1–19. Berkeley, CA: University of California Press.

Erlandson, Jon M., Todd J. Braje, Kristina M. Gill, and Michael H. Graham. 2015. Ecology of the Kelp Highway: Did Marine Resources Facilitate Human Dispersal from Northeast Asia to the Americas? *Journal of Island and Coastal Archaeology* 10(3): 392–411.

Etkin, Carol E. 1988. The Sechelt Indian Band: An Analysis of a New Form of Native Self Government. *The Canadian Journal of Native Studies* 8(1): 73–105.

Eurosion. 2004. Living with Coastal Erosion in Europe: Sediment and Space for Sustainability, PART I – Major findings and Policy Recommendations of the EUROSION project. Service contract B4-3301/2001/329175/MAR/B3, "Coastal erosion–Evaluation of the need for action," Directorate General Environment. Brussels: European Commission.

Evernden, Neil. 1995. Resourcism. In *Conservation and Environmentalism*, edited by R. Paehlke, p. 563. New York: Garland.

Eyring, Shaun, and Brian Goeken, eds. 2015. Cultural Resource Management and Planning for the Impacts of Climate Change. *George Wright Forum* 32(1): 35–6.

Ferris, Neal. 2002. When the Air Thins: The Rapid Rise of the Archaeological Consulting Industry in Ontario. *Journal of American Archaeology* 21: 53–88.

Fiala, Andrew. 2010. Nero's Fiddle: On Hope, Despair, and the Ecological Crisis. *Ethics and the Environment* 15(1): 51–68.

Flatman, Joe. 2009. Conserving Marine Cultural Heritage: Threats, Risks and Future Priorities. *Conservation and Management of Archaeological Sites* 11(1): 5–8.

Forbes, Jack D. 2008 [1978]. *Columbus and Other Cannibals: The Wétiko Disease of Exploitation, Imperialism, and Terrorism.* Revised Edition. New York: Seven Stories Press.

Foster, John Bellamy. 2009. *The Ecological Revolution: Making Peace with the Planet.* New York: Monthly Review Press.

Foster, John B., Brett Clark, and Richard York. 2010. *The Ecological Rift: Capitalism's War on the Earth.* New York: Monthly Review Press.

Fowler, Don D. 1987. Uses of the Past: Archaeology in the Service of the State. *American Antiquity* 52(2): 229–48.

French, Peter W. 2005. Coastal Zone Management. In *Encyclopedia of Coastal Science,* edited by M. Schwartz, pp. 313–19. Dordrecht: Springer.

Fulford, Michael, Timothy Champion, and Antony Long, eds. 1997. England's Coastal Heritage. English Heritage Archaeological Report 15. London: English Heritage and the RCHME.

Gelfenbaum, Guy, T. Mumford, J. Brennan, H. Case, M. Dethier, K. Fresh, F. Goetz, M. van Heeswijk, T. M. Leschine, M. Logsdon, D. Myers, J. Newton, H. Shipman, C. A. Simenstad, C. Tanner, and D. Woodson. 2006. Coastal Habitats in Puget Sound: A Research Plan in Support of the Puget Sound Nearshore Partnership. Puget Sound Nearshore Partnership Report No. 2006-1. Seattle, WA: US Geological Survey.

Georgia Basin-Puget Sound Ecosystem Indicators (GBPSEI). 2002. Transboundary Georgia Basin-Puget Sound Environmental Indicators Working Group. Georgia Basin Ecosystem Initiative Publication Number: EC/GB-01-034; Washington State Department of Ecology Publication Number: 02-01-002. Seattle, WA: Transboundary Georgia Basin-Puget Sound Environmental Indicators Working Group.

Georgia Basin-Puget Sound Ecosystem Indicators (GBPSEI). 2006. Georgia Basin-Puget Sound Ecosystem Indicators 2006. Vancouver, BC: Environment Canada; and Seattle, WA: US Environmental Protection Agency.

Giddens, Anthony. 1990. *The Consequences of Modernity.* Cambridge: Polity Press.

Glick, Patty, Jonathan Clough, and Brad Nunley. 2007. *Sea-level Rise and Coastal Habitats in the Pacific Northwest: An Analysis for Puget Sound, Southwestern Washington, and Northwestern Oregon.* Merrifield, VA: National Wildlife Federation.

Global Footprint Network. 2015. World Footprint. Electronic document, www.footprintnetwork.org/en/index.php/GFN/page/world_footprint/, accessed 08 November 2015.

Gnecco, Cristóbal. 2012. Europe and the People without Archaeology. In *European Archaeology Abroad: Global Settings, Comparative Perspectives,* edited by S. J. Van der Linde, M. H. van de Dries, N. Schlanger, and C. G. Slappendel, pp. 387–99. Leiden: Sidestone Press.

Goetz, Pauline. 2010. Preserving Arctic Archaeology in the 21st Century: Threats of Climate Change. MA thesis, Department of Anthropology, University of Waterloo, Ontario.

Goodhead, Tim, and David Johnson, eds. 1996. *Coastal Recreation Management: The Sustainable Development of Maritime Leisure*. New York: E & FN Spon.

Gosine, Andil, and Cheryl Teelucksingh. 2008. *Environmental Justice and Racism in Canada: An Introduction*. Toronto: Emond Montgomery.

Graber, Peter H. F. 2005. Coastal Boundaries. In *Encyclopedia of Coastal Science*, edited by M. Schwartz, pp. 246–51. Dordrecht: Springer.

Grinde, Donald A., and Bruce E. Johansen, eds. 1995. *Ecocide of Native America: Environmental Destruction of Indian Lands and Peoples*. Santa Fe, NM: Clear Light Publishers.

Grossman, Zoltán. 2012. No Longer the 'Miner's Canary.' In *Asserting Native Resilience: Pacific Rim Indigenous Nations Face the Climate Crisis*, edited by Z. Grossman and A. Parker, pp. 175–88. Corvallis, OR: Oregon State University Press.

Grossman, Zoltán, and Alan Parker, eds. 2012. *Asserting Native Resilience: Pacific Rim Indigenous Nations Face the Climate Crisis*. Corvallis, OR: Oregon State University Press.

Guinier, Lani, and Gerald Torres. 2002. *The Miner's Canary: Enlisting Race, Resisting Power, Transforming Democracy*. Cambridge: Harvard University Press.

Gurran, Nicole. 2008. The Turning Tide: Amenity Migration in Coastal Australia. *International Planning Studies* 13(4): 391–414.

Gurran, Nicole, Edward J. Blakely, and Caroline Squires. 2007. Governance Responses to Rapid Growth in Environmentally Sensitive Areas of Coastal Australia. *Coastal Management* 35(4): 445–65.

Hale, Sylvia M. 1990. *Controversies in Sociology*. Toronto: Copp Clark Pitman.

Hall, Thomas D., and James V. Fenelon. 2009. *Indigenous Peoples and Globalization: Resistance and Revitalization*. Boulder, CO: Paradigm.

Hamilakis, Yannis, and Phillip Duke, eds. 2007. *Archaeology and Capitalism: From Ethics to Politics*. Walnut Creek, CA: Left Coast Press.

Harford, Deborah, Nancy Olewiler, and John Richards. 2010. Climate Change Adaptation: Linkages with Social Policy. A Report by the Adaptation to Climate Change Team (ACT), from the Simon Fraser University (SFU) for the Government of Canada Policy Research Initiative. Ottawa: Government of Canada.

Harris, Anita. 2004. *Future Girl: Young Women in the Twenty-First Century*. New York: Routledge.

Harris, Cole. 1997. *The Resettlement of British Columbia: Essays on Colonialism and Geographical Change*. Vancouver: UBC Press.

Harrison, Rodney. 2013. *Heritage: Critical Approaches*. New York: Routledge.

Haslam, S. Alexander, and Stephen Reicher. 2012. Contesting the 'Nature' Of Conformity: What Milgram and Zimbardo's Studies Really Show. *PLoS Biology* 10(11): e1001426.

Havemann, Paul, and Helena Whall. 2002. The Miner's Canary: Indigenous Peoples and Sustainable Development in the Commonwealth. Commonwealth Policy Studies Unit Memorandum to Commonwealth Heads of Government attending the World Summit on Sustainable Development, Johannesburg, South Africa, 26 August–4 September 2002.

Hessing, Melody, and Michael Howlett. 1999. *Canadian Natural Resource and Environmental Policy*. Vancouver: UBC Press.

Hicks, Dan, and Laura McAtackney. 2007. Introduction: Landscapes as Standpoints. In *Envisioning Landscape: Situations and Standpoints in Archaeology and Heritage*, edited by D. Hicks, L. McAtackney and G. Fairclough, pp. 13–29. Walnut Creek, CA: Left Coast Press.

Homer-Dixon, Thomas. 2007. *The Upside of Down: Catastrophe, Creativity, and the Renewal of Civilization*. Toronto: Vintage Canada.

Horton, Benjamin P., Stefan Rahmstorf, Simon E. Engelhart, and Andrew C. Kemp. 2014. Expert Assessment of Sea-Level Rise by AD 2100 and AD 2300. *Quaternary Science Reviews* 84(15): 1–6.

Huppert, Daniel D., Amber Moore, and Karen Dyson. 2009. Impacts of Climate Change on the Coasts of Washington State. In *Washington Climate Change Impacts Assessment: Evaluating Washington's Future in a Changing Climate*, edited by J. S. Littell, M. McGuire Elsner, L. C. Whitely Binder, and A. K. Snove, pp. 285–309. Seattle, WA: Climate Impacts Group, University of Washington.

Hutchings, Richard M. 2010. Review of Avoiding Archaeological Disasters: A Risk Management Approach and Breaking Ground: The Lower Elwha Klallam Tribe and the Unearthing of Tse-whit-zen Village. *The Midden* 42(1–2): 24–7.

Hutchings, Richard M. 2011. A Response to The Midden's Forum on Media Representation and Cultural Resource Management in British Columbia. *The Midden* 43(3): 4–5.

Hutchings, Richard M. 2012a. Open Letter from Musqueam Indian Band Regarding Treatment of Musqueam Heritage and the 'Marpole Midden.' *The Midden* 44(2): 6–7.

Hutchings, Richard M. 2012b. "Kennewick Man" Neither Native American nor Indigenous to Columbia Valley says Federal Archaeologist. *The Midden* 44(3/4): 3–4.

Hutchings, Richard M. 2013. "Hard Times Bring Hard Questions"—Is Archaeology Pro-Development? Is it Classist? Colonialist? Imperialist? Racist? Plenary paper presented at the Canadian Archaeological Association Conference, Whistler, BC, 15–19 May 2013. Electronic document, http://viu.academia.edu/RichHutchings, accessed 04 March 2014.

Hutchings, Richard M. 2014. Lies, Damn Lies, and CRM: A Response to the Austin Declaration on the "Excavation of Archaeological Material in the Popular Media." *Institute for Critical Heritage Studies Bulletin 2014-3*. Nanaimo, BC: Institute for Critical Heritage and Tourism.

Hutchings, Richard M. 2016. On Commercial, Compliance, or Contract Archaeology. *The Midden* 46(1–2): 28.

Hutchings, Richard M., and Marina La Salle. 2013. Five Thoughts on Commercial Archaeology. *Institute for Critical Heritage Studies Bulletin 2013–1*. Nanaimo, BC: Institute for Critical Heritage and Tourism.

Hutchings, Richard M., and Marina La Salle. 2014. Teaching Anti-Colonial Archaeology. *Archaeologies: Journal of the World Archaeological Congress* 10(1): 7, 27–69.

Hutchings, Richard M., and Marina La Salle. 2015a. Archaeology as Disaster Capitalism. *International Journal of Historical Archaeology* 19(4): 699–720.

Hutchings, Richard M., and Marina La Salle. 2015b. Arqueologia como Capitalismo do Desastre. *Revista de Arqueologia da Sociedade de Arqueologia Brasileira* 28(2): 20–44.

Hutchings, Richard M., and Marina La Salle. 2015c. Why Archaeologists Misrepresent Their Practice—A Northern American Perspective. *Journal of Contemporary Archaeology* 2(2): S11–S17.

Ingold, Tim. 2011. *Being Alive: Essays on Movement, Knowledge and Description*. New York: Routledge.

International Council on Monuments and Sites (ICOMOS). 2006. *Underwater Cultural Heritage at Risk: Managing Natural and Human Impacts*. Paris: ICOMOS.

International Council on Monuments and Sites (ICOMOS). 2010. *World Report on Monuments and Sites in Danger 2008–2010*. Paris: ICOMOS.

International Geosphere–Biosphere Project. 2015. Great Acceleration. Electronic document, www.igbp.net/globalchange/greatacceleration.4.1b8ae20512db692f2a680001630.html, accessed 18 June 2015.

Jackson, Jeremy, and Karen Alexander. 2011. Introduction: The Importance of Shifting Baselines. In *Shifting Baselines: The Past and the Future of Ocean Fisheries*, edited by J. B. C. Jackson, K. E. Alexander, and E. Sala, pp. 1–7. Washington, Covelo, and London: Island Press.

Jaimes, M. Annette, ed. 1992. *The State of Native America: Genocide, Colonization, and Resistance*. Boston, MA: South End Press.

Jefferson, Thomas. 1801. [Letter] To the Governor of Virginia [James Monroe], Washington, Nov. 24, 1801. In *The Works of Jefferson IX*, edited by P. L. Ford, pp. 315–19. New York and London: G. P. Putnam's Sons.

Jensen, Derrick, and Aric McBay. 2009. *What We Leave Behind*. New York: Seven Stories Press.

Johannessen, Jim, and Andrea MacLennan. 2007. Beaches and Bluffs of Puget Sound and the Northern Straits. Puget Sound Nearshore Partnership. Puget Sound Nearshore Partnership Report No. 2007-04. Seattle, WA: US Army Corps of Engineers, Seattle District.

Joint Programming Institute (JPI). 2009. *Cultural Heritage, Climate Change and Security*. Rome: Joint Programming Institute on Cultural Heritage.

Jones, Benjamin M., Kenneth M. Hinkel, Christopher D. Arp, and Wendy R. Eisner. 2008. Modern Erosion Rates and Loss of Coastal Features and Sites, Beaufort Sea Coastline, Alaska. *Arctic* 61(4): 361–72.

Kahan, Dan M., Hans Jenkins-Smith, and Donald Braman. 2011. Cultural Cognition of Scientific Consensus. *Journal of Risk Research* 14(2): 147–74.

Kerr Wood Leidal. 2011. Coastal Floodplain Mapping – Guidelines and Specifications. Report prepared by Kerr Wood Leidal Consulting Engineers for the Province of British Columbia and Natural Resources Canada, Ministry of Forests, Lands and Resources Operations. Ottawa: Government of Canada.

King, Thomas F. 1998. *Cultural Resource Laws and Practice*. Walnut Creek, CA: AltaMira Press.

King, Thomas F. 2008. Cultural Resource Management and Historic Properties: Learning from the Errors of the United States. Electronic document, http://crmplus.blogspot.com/2010/04/dont-do-what-we-did-small-critique-of.html, accessed 08 August 2011.

King, Thomas F. 2009. *Whitewashing the Destruction of Our Natural and Cultural Heritage*. Walnut Creek, CA: Left Coast Press.

Klein, Naomi. 2015. This Changes Everything. *The Guardian* 08 March 2015. Electronic document, www.theguardian.com/environment/2015/mar/08/how-will-everything-change-under-climate-change, accessed 04 April 2015.

La Salle, Marina. 2010. Community Collaboration and Other Good Intentions. *Archaeologies: The Journal of the World Archaeological Congress* 6(3): 401–22.

La Salle, Marina. 2013. 'Capital-C' Consultation: Community, Capitalism and Colonialism. *New Proposals: Journal of Marxism and Interdisciplinary Inquiry* 6(1–2): 72–88.

La Salle, Marina. 2014a. Escape into Nature: The Ideology of Pacific Spirit Regional Park. PhD dissertation. Department of Anthropology, University of British Columbia, Vancouver.

La Salle, Marina. 2014b. The Trouble with 'Co-.' *Institute for Critical Heritage Studies Bulletin 2014-1*. Nanaimo, BC: Institute for Critical Heritage and Tourism.

La Salle, Marina, and Richard M. Hutchings. 2012. Commercial Archaeology in British Columbia. *The Midden* 44(2): 8–16.

La Salle, Marina, and Richard M. Hutchings. 2015. Review of Transforming Archaeology: Activist Practices and Prospects. *Canadian Journal of Archaeology* 39(2): 359–62.

La Salle, Marina, and Richard M. Hutchings. 2016. What Makes Us Squirm: A Response to Community-Oriented Archaeology. *Canadian Journal of Archaeology* 40(1): 164–80.

Landry, Craig E. 2011. Coastal Erosion as a Natural Resource Management Problem: An Economic Perspective. *Coastal Management* 39(3): 259–81.

Lertzman, Ken. 2009. The Paradigm of Management, Management Systems, and Resource Stewardship. *Journal of Ethnobiology* 29(2): 339–58.

Lyons, Natasha, Ian Cameron, Tanya Hoffmann, and Debbie Miller. 2012. Many Shades of Grey: Dispelling Some Myths About the Nature and Status of CRM in British Columbia, a Response to La Salle and Hutchings. *The Midden* 44(3/4): 6–8.

McFadden, Loraine. 2007. Governing Coastal Spaces: The Case of Disappearing Science in Integrated Coastal Zone Management. *Coastal Management* 35(4): 429–43.

McGinnis, Michael V., and Christina E. McGinnis. 2011. Adapting to Climate Impacts in California: The Importance of Civic Science in Local Coastal Planning. *Coastal Management* 39(3): 225–41.

Mack, John. 2011. *The Sea: A Cultural History*. London: Reaktion.

McLay, Eric. 2011. 'Site Alteration': The Demolition of British Columbia's Archaeological Heritage. *The Midden* 43(4): 3–7.

McLay, Eric, Kelly Bannister, Lea Joe, Brian Thom, and George Nicholas. 2008. *A'lhut tu tet Sulhweenst* [Respecting the Ancestors]: Understanding Hul'qumi'num Heritage Laws and Concerns for the Protection of Archaeological Heritage. In *First Nations Cultural Heritage and Law: Case Studies, Voices, and Perspectives*, edited by C. Bell and V. Napoleon, pp. 150–202. Vancouver: UBC Press.

McNiven, Ian J., and Lynette Russell. 2005. *Appropriated Pasts: Indigenous Peoples and the Colonial Culture of Archaeology*. Lanham, MD: AltaMira Press.

Magdoff, Fred, and John Bellamy Foster. 2011. *What Every Environmentalist Needs to Know About Capitalism*. New York: Monthly Review Press.

Mallette, Sebastien. 2009. Foucault for the Next Century: Eco-Governmentality. In *A Foucault for the 21st Century*, edited by S. Binkley, pp. 221–39. Cambridge: Cambridge Scholars Publishing.

Mander, Jerry. 1992. *In the Absence of the Sacred: The Failure of Technology and the Survival of the Indian Nations*. San Francisco, CA: Sierra Club Books.

Manson, Gavin K. 2005. On the Coastal Populations of Canada and the World. Proceedings of the 12th Canadian Coastal Conference, Dartmouth, Nova Scotia, 6–9 November 2005.

Mapes, Lynda V. 2009. *Breaking Ground: The Lower Elwha Klallam Tribe and the Unearthing of Tse-whit-zen Village*. Seattle, WA: University of Washington Press.

Markey, Nola M. 2001. "Data Gathering Dust": An Analysis of Traditional Use Studies Conducted within Aboriginal Communities in British Columbia. MA thesis, Department of Archaeology, Simon Fraser University, Burnaby, British Columbia.

Maud, Ralph, ed. 1978. *The Salish People—The Local Contributions of Charles Hill-Tout, Volume IV: The Sechelt and the South-Eastern Tribes of Vancouver Island*. Vancouver: Talonbooks.

Meighan, Clement, and Hal Eberhart. 1953. Archaeological Resources of San Nicolas Island, California. *American Antiquity* 19(2): 109–25.

Menzies, Charles R. ed. 2006. *Traditional Ecological Knowledge and Natural Resource Management*. Lincoln: University of Nebraska Press.

Merchant, Carolyn. 1992. *Radical Ecology: The Search for a Livable World*. New York and London: Routledge.

Merchant, Peter. 2004. Archaeological Inventory of Sixteen Kilometres of Foreshore along the Coastline of Pender Harbour within the Territory of the *shíshálh* Nation, Southwestern BC. Report on file with the *shíshálh* First Nation. Sechelt, BC: *shíshálh* First Nation.

Merchant, Peter. 2010. Re-Inflicting the Past: Reflections on Personal Experiences in Collaboration. *Archaeological Review from Cambridge* 26(2): 45–57.

Merchant, Peter. 2012. *shíshálh* Responses to the Colonial Conflict (1791–present): Resilience in the Face of Disease, Missionaries and Colonization. MA thesis, Department of Anthropology, University of British Columbia, Vancouver.

Millennium Ecosystem Assessment. 2005a. *Ecosystems and Human Well-being: Synthesis*. Washington, DC: Island Press.

Millennium Ecosystem Assessment. 2005b. *Ecosystems and Human Well-Being: Wetlands and Water – Synthesis*. Washington, DC: World Resources Institute.

Milner, Nicky. 2012. Destructive Events and the Impact of Climate Change on Stone Age Coastal Archaeology in North West Europe: Past, Present and Future. *Journal of Coastal Conservation* 16(2): 223–31.

Mitchell, Bruce, and W. R. D. Sewell. 1981. The Emerging Scene. In *Canadian Resource Policies*, edited by B. Mitchell and W. R. D. Sewell, pp. 1–24. Agincourt, ON: Methuen.

Mitchell, David J. 1983. *W. A. C. Bennett and the Rise of British Columbia*. Vancouver and Toronto: Douglas and McIntyre.

Mitchell, Peter. 2008. Practising Archaeology at a Time of Climatic Catastrophe. *Antiquity* 82(318): 1093–103.

Moghaddam, Fathali M. 1997. *The Specialized Society: The Plight of the Individual in an Age of Individualism*. Westport, CT: Praeger.

Moghaddam, Fathali M. 2010. *The New Global Insecurity: How Terrorism, Environmental Collapse, Economic Inequalities, and Resource Shortages Are Changing Our World*. Santa Barbara, CA: Praeger Security International.

Monbiot, George. 2015. We're Treating Soil Like Dirt. It's a Fatal Mistake, As Our Lives Depend on It. *The Guardian*, 25 March 2015. Electronic document, www.theguardian.com/commentisfree/2015/mar/25/treating-soil-like-dirt-fatal-mistake-human-life, accessed 18 June 2015.

Moss, Madonna L., and Jon M. Erlandson. 2008. Native American Archaeological Sites of the Oregon Coast: The Historic Context for the Nomination to the National Register of Historic Places. In *Dunes, Headlands, Estuaries, and Rivers: Current Archaeological Research on the Oregon Coast*, edited by G. L. Tasa and B. L. O'Neill, pp. 1–36. Eugene, OR: Association of Oregon Archaeologists.

Muckle, Robert. 2013. Perceived Conflicts in Archaeology-for-Profit. *Anthropology News*, 21 April 2014. Electronic document, www.anthropology-news.org/index.php/2013/09/30/perceived-conflicts-in-archaeology-for-profit/, accessed 28 April 2014.

Murphy, Peter, David Thackray, and Ed Wilson. 2009. Coastal Heritage and Climate Change in England: Assessing Threats and Priorities. *Conservation and Management of Sites* 11(1): 9–15.

Myers, Doug. 2010. Shoreline Development on Puget Sound. In *Puget Sound Shorelines and the Impacts of Armoring—Proceedings of a State of the Science Workshop, May 2009*, edited by H. Shipman, M. N. Dethier, G. Gelfenbaum, K. L. Fresh, and R. S. Dinicola, pp. 43–8. US Geological Survey Scientific Investigations Report 2010-5254. Reston, VA: US Geological Survey.

Myers, Norman. 1997. Environmental Refugees. *Population and Environment* 19(2):167–82.
Nash, Roderick F. 2014. *Wilderness and the American Mind*. 5th Edition. New Haven, CT: Yale University Press.
National Oceanic and Atmospheric Administration. 2015. Global Summary Information - December 2014. Electronic document, www.ncdc.noaa.gov/sotc/summary-info/global/201412, accessed 02 February 2015.
National Round Table on the Environment and the Economy. 2011. *Paying the Price: The Economic Impacts of Climate Change for Canada*. Ottawa: National Round Table on the Environment and the Economy.
Neal, William J., David M. Bush, and Orrin H. Pilkey. 2005. Managed Retreat. *In Encyclopedia of Coastal Science*, edited by M. Schwartz, pp. 602–6. Dordrecht: Springer.
Ng'ang'a, Sam Macharia. 2006. Extending Land Management Approaches to Coastal and Oceans Management: A Framework for Evaluating the Role of Tenure Information in Canadian Marine Protected Areas. Technical Report No. 243, Department of Geodesy and Geomatics Engineering. Fredericton, NB: University of New Brunswick.
Nicholas, George P. 2006. Decolonizing the Archaeological Landscape: The Practice and Politics of Archaeology in British Columbia. *American Indian Quarterly* 30(3): 350–80.
Nicholas, George P., and Julie Hollowell. 2007. Ethical Challenges to a Postcolonial Archaeology: The Legacy of Scientific Colonialism. *In Archaeology and Capitalism: From Ethics to Politics*, edited by Y. Hamilakis and P. Duke, pp. 2, 59–82. Walnut Creek, CA: Left Coast Press.
Nicholls, Robert J., P. P. Wong, V. R. Burkett, J. O. Codignotto, J. E. Hay, R. F. McLean, S. Ragoonaden, and C. D. Woodroffe. 2007. Coastal Systems and Low-Lying Areas. In *Climate Change 2007: Impacts, Adaptation and Vulnerability. Contribution of Working Group II to the Fourth Assessment Report of the Intergovernmental Panel on Climate Change*, edited by M. L. Parry, O. Canziani, J. Palutikof, P. van der Linden, and C. Hanson, pp. 315–56. Cambridge: Cambridge University Press.
Nixon, Rob. 2006–2007. Slow Violence, Gender, and the Environmentalism of the Poor. *Journal of Commonwealth and Postcolonial Studies* 13/14: 14–37.
Nixon, Rob. 2011a. *Slow Violence and the Environmentalism of the Poor*. Harvard, MA: Harvard University Press.
Nixon, Rob. 2011b. Slow Violence: Literary and Postcolonial Studies Have Ignored the Environmentalism that Often Only the Poor Can See. *The Chronicle of Higher Education*, 02 June 2011. Electronic document, http://chronicle.com/article/Slow-Violence/127968/, accessed 31 January 2014.
Nutley, David M. 2005. Surviving Inundation: An Examination of Environmental Factors Influencing the Survival of Inundated Indigenous Sites in Australia. MA thesis, Department of Archaeology, Flinders University of South Australia, Australia.
O'Connor, Elaine. 2014. Dusting Off Archaeology in B.C.: How the Resource and Development Boom is Enabling a New Generation of Archaeologists to Dig for a Living. *The Province*, 13 April 2014: A8.
Oertel, George F. 2005. Coasts, Coastlines, Shores, and Shorelines. In *Encyclopedia of Coastal Science*, edited by M. Schwartz, pp. 323–7. Dordrecht: Springer.
Ommer, Rosemary, with the Coasts Under Stress Research Project Team. 2007. *Coasts Under Stress: Restructuring and Social-Ecological Health*. Montreal and Kingston: McGill-Queen's University Press.
Orwell, George. 1949. *Nineteen Eighty-Four*. New York: Signet Classics.
O'Sullivan, John. 1845. Annexation. *United States Magazine and Democratic Review* 17(1): 5–10.

Parker, Alan. 2012. Recommendations to Native Government Leadership. In *Asserting Native Resilience: Pacific Rim Indigenous Nations Face the Climate Crisis*, edited by Z. Grossman and A. Parker, pp. 189–92. Corvallis. OR: Oregon State University Press.

Parker, Alan, and Zoltán Grossman. 2012. Introduction. In *Asserting Native Resilience: Pacific Rim Indigenous Nations Face the Climate Crisis*, edited by Z. Grossman and A. Parker, pp. 13–19. Corvallis, OR: Oregon State University Press.

Perry, Jim. 2011. World Heritage Hot Spots: A Global Model Identifies the 16 Natural Heritage Properties on the World Heritage List Most at Risk from Climate Change. *International Journal of Heritage Studies* 17(5): 426–41.

Peterson, Lester R. 1962. *The Gibson's Landing Story*. Toronto: P. Martin Books.

Peterson, Lester R. 1990. *The Story of the Sechelt Nation*. Madeira Park, BC: Harbour Publishing.

Pilkey, Orrin H., and Rob Young. 2009. *The Rising Sea*. Washington, Covelo, and London: Island Press.

Pope John Paul II. 1990. Peace with God the Creator, Peace with All of Creation. Message of His Holiness Pope John Paul II for the Celebration of the World Day of Peace, 01 January 1990. Electronic document, https://w2.vatican.va/content/john-paul-ii/en/messages/peace/documents/hf_jp-ii_mes_19891208_xxiii-world-day-for-peace.html, accessed 16 June 2015.

Prosper, Lisa. 2007. Wherein Lies the Heritage Value? Rethinking the Heritage Value of Cultural Landscapes from an Aboriginal Perspective. *The George Wright Forum* 24(2): 117–24.

Puget Sound Partnership (PSP). 2011. *Puget Sound Science Update*. Tacoma, WA: Puget Sound Partnership.

Quinn, Timothy. 2010. An Environmental and Historical Overview of the Puget Sound Ecosystem. In *Puget Sound Shorelines and the Impacts of Armoring—Proceedings of a State of the Science Workshop, May 2009*, edited by H. Shipman, M. N. Dethier, G. Gelfenbaum, K. L. Fresh, and R. S. Dinicola, pp. 11–17. US Geological Survey Scientific Investigations Report 2010-5254. Reston, VA: US Geological Survey.

Raban, Jonathan. 1992. *The Oxford Book of the Sea*. New York: Oxford University Press.

Reeder, Leslie A., Torben C. Rick, and Jon M. Erlandson. 2012. Our Disappearing Past: A GIS Analysis of the Vulnerability of Coastal Archaeological Resources in California's Santa Barbara Channel Region. *Journal of Coast Conservation* 16(2): 187–97.

Regan, Paulette. 2010. *Unsettling the Settler Within: Indian Residential Schools, Truth Telling, and Reconciliation in Canada*. Vancouver: UBC Press.

Reicher, Stephen, and S. Alexander Haslam. 2006. Rethinking the Psychology of Tyranny: The BBC Prison Study. *British Journal of Social Psychology* 45(1): 1–40.

Reid, Walter V., and Marc C. Trexler. 1991. *Drowning the National Heritage: Climate Change and U.S. Coastal Biodiversity*. Washington, DC: World Resources Institute.

Rick, Torben C., and Scott M. Fitzpatrick. 2012. Archaeology and Coastal Conservation. *Journal of Coastal Conservation* 16(2): 135–6.

Rick, Torben C., Jon M. Erlandson, and René L. Vellanoweth. 2006. Taphonomy and Site Formation on California's Channel Islands. *Geoarchaeology* 21(6): 567–89.

Ridgeway, Sharon J., and Peter J. Jacques. 2014. *The Power of the Talking Stick: Indigenous Politics and the World Ecological Crisis*. Boulder, CO: Paradigm.

Robbins, Richard H. 2005. *Global Problems and the Culture of Capitalism*. Boston: Allyn & Bacon.

Rogers, Raymond A. 1994. *Nature and the Crisis of Modernity: A Critique of Contemporary Discourse on Managing the Earth*. Montreal: Black Rose Books.

Rogers, Raymond A. 1995. *The Oceans Are Emptying: Fish Wars and Sustainability*. Montreal, New York, and London: Black Rose Books.

Rogers, Raymond A. 1998. *Solving History: The Challenge of Environmental Activism*. Montreal: Black Rose Books.

Roosevelt, Franklin D. 1937. Letter to all State Governors on a Uniform Soil Conservation Law, February 26, 1937. Electronic document, www.presidency.ucsb.edu/ws/?pid=15373, accessed 12 March 2014.

Rose, Deborah. 2005. An Indigenous Philosophical Ecology: Situating the Human. *The Australian Journal of Anthropology* 16(3): 294–305.

Ross, Anne, Kathleen P. Sherman, Jeffery G. Snodgrass, Henry D. Delcore, and Richard Sherman. 2011. *Indigenous People and the Collaborative Stewardship of Nature*. Walnut Creek. CA: Left Coast Press.

Roy, Susan. 2010. *These Mysterious People: Shaping History and Archaeology in a Northwest Coast Community*. Montreal and Kingston: McGill-Queen's University Press.

Rozen, David Lewis. 1985. Place-names of the Island Halkomelem Indian People. MA thesis, Department of Anthropology, University of British Columbia, Vancouver.

Safford, Thomas G., and Lawrence C. Hamilton. 2012. Demographic Change and Shifting Views about Marine Resources and the Coastal Environment in Downeast Maine. *Population and Environment* 33(4): 284–303.

Schiffer, Michael B. 1987. *Formation Processes of the Archaeological Record*. Salt Lake City, UT: University of Utah Press.

Schrecker, Ted. 2001. Using Science in Environmental Policy: Can Canada Do Better? In *Governing the Environment: Persistent Challenges, Uncertain Innovations*, edited by E. A. Parson, pp. 31–72. Toronto: University of Toronto Press.

Schwartz, Maurice, ed. 2005. *Encyclopedia of Coastal Science*. Dordrecht: Springer.

Scott, David B. 2005. Coastal Change, Rapid. In *Encyclopedia of Coastal Science*, edited by M. Schwartz, pp. 253–5. Dordrecht: Springer.

Scott, James C. 1998. *Seeing Like a State: How Certain Schemes to Improve the Human Condition Have Failed*. New Haven, CT: Yale University Press.

Sechelt Indian Band (SIB). 2007. *lil xemit tems swiya nelh mes stutula:* A Strategic Land Use Plan for the *shíshálh* Nation. Sechelt, BC: *shíshálh* First Nation.

Sechelt Indian Band (SIB). 2013. *shíshálh* Nation Lands and Resources Decision-Making Policy. Sechelt, BC: *shíshálh* First Nation.

Shanks, Michael, and Christopher Tilley. 1987. *Re-Constructing Archaeology: Theory and Practice*. Cambridge: Cambridge University Press.

Shaw, J., R. B. Taylor, D. L. Forbes, M. H. Ruz, and S. Solomon. 1998. Sensitivity of the Coasts of Canada to Sea-Level Rise. *Geological Survey of Canada Bulletin 505*. Ottawa: Environment Canada.

Shearing, Susan. 2007. Here Today, Gone Tomorrow—Climate Change and World Heritage. Macquarie Law Working Paper No. 2007-11. Macquarie Park, Australia: Macquarie Law School.

Shipman, Hugh. 2004. Coastal Bluffs and Sea Cliffs on Puget Sound, Washington. In *Formation, Evolution, and Stability of Coastal Cliffs—Status and Trends*, edited by M. Hampton and G. Griggs, pp. 81–94. US Geological Survey Professional Paper 1693. Denver, CO: US Geological Survey.

Shipman, Hugh. 2008. A Geomorphic Classification of Puget Sound Nearshore Landforms. Puget Sound Nearshore Partnership Report No. 2008-01. Seattle, WA: US Army Corps of Engineers, Seattle District.

Shipman, Hugh. 2010. The Geomorphic Setting of Puget Sound—Implications for Shoreline Erosion and the Impacts of Erosion Control Structures. In *Puget Sound Shorelines and the Impacts of Armoring—Proceedings of a State of the Science Workshop, May 2009*, edited by H. Shipman, M. N. Dethier, G. Gelfenbaum, K. L. Fresh, and R. S. Dinicola, pp. 19–34. US Geological Survey Scientific Investigations Report 2010-5254. Reston, VA: US Geological Survey.

Shipman, Hugh., Megan N. Dethier, Guy Gelfenbaum, Kurt L. Fresh, and Richard S. Dinicola, eds. 2010a. Puget Sound Shorelines and the Impacts of Armoring—Proceedings of a State of the Science Workshop, May 2009. US Geological Survey Scientific Investigations Report 2010–5254. Reston, VA: US Geological Survey.

Shipman, Hugh, Guy Gelfenbaum, Megan N. Dethier, and Kurt L. Fresh. 2010b. Puget Sound Shoreline Armoring: State of the Science Workshop. In *Puget Sound Shorelines and the Impacts of Armoring—Proceedings of a State of the Science Workshop, May 2009*, edited by H. Shipman, M. N. Dethier, G. Gelfenbaum, K. L. Fresh, and R. S. Dinicola, pp. 1–8. US Geological Survey Scientific Investigations Report 2010-5254. Reston, VA: US Geological Survey.

Smith, Angèle. 2008. Landscapes of Clearance: Archaeological and Anthropological Perspectives. In *Landscapes of Clearance*, edited by A. Smith and A. Gazin-Schwartz, pp. 13–24. Walnut Creek, CA: Left Coast Press.

Smith, Laurajane. 2001. Archaeology and the Governance of Material Culture: A Case Study from South-Eastern Australia. *Norwegian Archaeological Review* 34(2): 97–105.

Smith, Laurajane. 2004. *Archaeological Theory and the Politics of Cultural Heritage*. New York: Routledge.

Smith, Laurajane. 2006. *Uses of Heritage*. New York: Routledge.

Spence, Mark David. 1999. *Dispossessing the Wilderness: Indian Removal and the Making of the National Parks*. New York: Oxford University Press.

Springer, Simon. 2010. Neoliberalism and Geography: Expansions, Variegations, Formations. *Geography Compass* 4(8): 1025–38.

Spurling, Brian E. 1986. Archaeological Resource Management in Western Canada: A Policy Science Approach. PhD dissertation, Department of Archaeology, Simon Fraser University, Burnaby, BC.

Spurling, Brian E. 1988. Archaeology and the Policy Sciences. *Canadian Journal of Archaeology* 12: 65–85.

Stanton, Elizabeth A., Marion Davis, and Amanda Fencl. 2010. Costing Climate Impacts and Adaptation: A Canadian Study on Coastal Zones. Report Commissioned by the National Roundtable on the Environment and the Economy. Somerville, MA: Stockholm Environment Institute.

Stapp, Darby, and Julia Longenecker. 2009. *Avoiding Archaeological Disasters: A Risk Management Approach*. Walnut Creek, CA: Left Coast Press.

State of Washington Office of Financial Management. 2010. Forecast of the State Population: November 2010 Forecast. Olympia, WA: Washington State Office of Financial Management.

Steeves, Paulette. 2015. Academia, Archaeology, CRM, and Tribal Historic Preservation. *Archaeologies: Journal of the World Archaeological Congress* 11(1): 121–41.

Steffen, Will, Wendy Broadgate, Lisa Deutsch, Owen Gaffney, and Cornelia Ludwig. 2015. The Trajectory of the Anthropocene: The Great Acceleration. *The Anthropocene Review* 2(1): 81–98.

Steffen, Will, A. Sanderson, P. D. Tyson, J. Jäger, P. A. Matson, B. Moore III, F. Oldfield, K. Richardson, H. J. Schellnhuber, B. L. Turner, and R. J. Wasson. 2004. *Global Change and the Earth System: A Planet Under Pressure.* Berlin: Heidelberg and New York: Springer-Verlag.

Stein, Julie K. 2001. *Deciphering a Shell Midden.* New York: Harcourt Brace Jovanovich.

Sunshine Coast Credit Union. 2011. The Future Business Environment of the Sunshine Coast 2011–2021. Sechelt, BC: Economics Department, Central 1 Credit Union.

Sutton, Mary-Jean, Jillian Huntley, and Barry Anderson. 2013. "All Our Sites are of High Significance": Reflections from Recent Work in the Hunter Valley—Archaeological and Indigenous Perspective. *Journal of the Australian Association of Consulting Archaeologists* 1: 1–15.

Task Force. 2005. Seattle/King County Task Force on Maritime Heritage: Final Report, December 6, 2005. 4Culture, Association of King County Historical Organizations, Historic Seattle, National Park Service Pacific West Region, and Washington Trust for Historic Preservation. Seattle, WA: Task Force on Maritime Heritage.

Taylor, Kathy, Hilary Culverwell, and John Cambalik. 2005. Educating Shoreline Landowners: Examples from King, Whatcom, Kitsap, Jefferson, Mason, and Pierce Counties; A Perspective on Approaches and Effectiveness in Eliciting on-the-Ground Change. Proceedings of the 2005 Puget Sound Georgia Basin Research Conference, Seattle, WA, 29–31 March 2005.

The National Trust. 2005. *Shifting Shores: Living with a Moving Shoreline.* Exeter, UK: The National Trust.

Theodorson, George A., and Achilles G. Theodorson. 1969. *A Modern Dictionary of Sociology.* New York and London: Harper and Row.

Thorburne, Craig Carpenter. 2000. *Kau Kuat, Kau Pinter, Kau Punya* (You're Strong, You're Clever, It's Yours): Changing Coastal Resource Management Institutions and Practice in the Kei Islands, Eastern Indonesia. PhD dissertation, Department of Geography, University of California, Los Angeles.

Tomalin, David. 1998. Bargaining with Nature: Considering the Sustainability of Archaeological Sites in the Dynamic Environment of the Intertidal Zone. In *Preserving Archaeological Remains In-Situ: Proceedings of the Conference of 1st–3rd April 1996*, edited by M. Cornfield, P. Hinton, T. Nixon, and M. Pollard, pp. 144–86. London and Bradford: Museum of London/University of Bradford.

Trigger, Bruce. 1989. *A History of Archaeological Thought.* Cambridge: Cambridge University Press.

Turner, Nancy, Iain J. Davidson-Hunt, and Michael O'Flaherty. 2003. Living on the Edge: Ecological and Cultural Edges as Sources of Diversity for Social-Ecological Resilience. *Human Ecology* 31(3): 439–61.

Tylor, Edward B. 1974 [1871]. *Primitive Culture: Researches into the Development of Mythology, Philosophy, Religion, Art, and Custom.* New York: Gordon Press.

Union of British Columbia Indian Chiefs (UBCIC). 2013. *First Nations Heritage Planning Toolkit.* Vancouver: Union of BC Indian Chiefs.

United Nations (UN). 2008. *United Nations Declaration on the Rights of Indigenous Peoples.* New York: United Nations General Assembly.

United Nations Educational, Scientific and Cultural Organization (UNESCO). 1972. *Convention Concerning the Protection of the World Cultural and Natural Heritage.* Paris: UNESCO.

United Nations Educational, Scientific and Cultural Organization (UNESCO). 2003. *Convention for the Safeguarding of the Intangible Cultural Heritage.* Paris: UNESCO.

United Nations Educational, Scientific and Cultural Organization (UNESCO). 2007a. Climate Change and World Heritage: Report on Predicting and Managing the Impacts of Climate Change on World Heritage and Strategy to Assist States Parties to Implement Appropriate Management Responses. World Heritage Reports 22. Paris: UNESCO.

United Nations Educational, Scientific and Cultural Organization (UNESCO). 2007b. *Case Studies on Climate Change and World Heritage*. Paris: UNESCO.

United Nations Educational, Scientific and Cultural Organization (UNESCO). 2008. Policy Document on the Impacts of Climate Change on World Heritage Properties. Paris: UNESCO.

United Nations Educational, Scientific and Cultural Organization (UNESCO). 2009. *Case Studies on Climate Change and World Heritage*. Paris: UNESCO.

United Nations Educational, Scientific and Cultural Organization (UNESCO). 2010. *Managing Disaster Risks for World Heritage*. Paris: UNESCO.

United States Census Bureau. 2005. Table A1: Interim Projections of the Total Population for the United States and States: April 1, 2000 to July 1, 2030. US Census Bureau, Population Division, Interim State Population Projections, 2005. Washington, DC: US Department of Commerce.

United States Environmental Protection Agency/Environment Canada. 2013. Salish Sea Transboundary Ecosystem Indicators: Executive Summary. Washington, DC and Ottawa: United States Environmental Protection Agency and Environment Canada.

United States Geological Survey (USGS). 1997. Changes in Water Level. Electronic document, http://pubs.usgs.gov/circ/c1075/level.html, accessed 08 August 2013.

Vidal, John. 2015. Explosive Intervention by Pope Francis Set to Transform Climate Change Debate. *The Guardian* 13 June. Electronic document, www.theguardian.com/world/2015/jun/13/pope-francis-intervention-transforms-climate-change-debate, accessed 15 June 2015.

Walcott, Derek. 1979. *The Star-Apple Kingdom*. New York: Farrar, Straus and Giroux.

Washington State Department of Archaeology and Historic Preservation (DAHP). 2009. Washington State National Maritime Heritage Area Feasibility Study. Olympia, WA: Washington State Department of Archaeology and Historic Preservation.

Washington State Department of Archaeology and Historic Preservation (DAHP). 2010. Washington State National Maritime Heritage Area Feasibility Study for Designation as a National Heritage Area. Olympia, WA: Washington State Department of Archaeology and Historic Preservation.

Waters, Michael R. 1992. *Principles of Geoarchaeology: A North American Perspective*. Tucson, AZ: University of Arizona Press.

Waterton, Emma, and Laurajane Smith, eds. 2009. *Taking Archaeology Out of Heritage*. Newcastle upon Tyne: Cambridge Scholars.

Watson, Christopher S., Neil J. White, John A. Church, Matt A. King, Reed J. Burgette, and Benoit Legresy. 2015. Unabated Global Mean Sea-Level Rise over the Satellite Altimeter Era. *Nature Climate Change* 5: 565–8.

Weber, Max. 1964 [1922]. *Economy and Society: An Outline of Interpretive Sociology*. Trans. G. Roth and G. Wittich. New York: Bedminster Press.

Weil, Simone. 1952. *The Need for Roots*. New York and London: Routledge.

Welch, John R., and Neal Ferris. 2014. "We Have Met the Enemy and It Is Us": Transforming Archaeology through Sustainable Design. In *Transforming Archaeology: Activist Practices and Prospects*, edited by S. Atalay, L. R. Clauss, R. H. McGuire, and J. R. Welch, pp. 91–113. Walnut Creek, CA: Left Coast Press.

Welch, John R., Eric McLay, Michael Klassen, Fred Foster, and Robert Muir. 2011. A Database of Unauthorized Heritage Site Alterations. *The Midden* 43(1): 2–3.

Wells, Lisa E. 2001. Archaeological Sediments in Coastal Environments. In *Sediments in Archaeological Context*, edited by J. K. Stein and W. Farrand, pp. 149–82. Salt Lake City, UT: The University of Utah Press.

Wendorf, Fred. 1963. Archaeology and Private Enterprise: A Need for Action. *American Antiquity* 28(3): 286–8.

West, Robert Gerard. 1995. Saving and Naming the Garbage: Charles E. Borden and the Making of B.C. Prehistory. MA thesis, Department of History, University of British Columbia, Vancouver.

Westerdahl, Christer. 1992. The Maritime Cultural Landscape. *The International Journal of Nautical Archaeology* 21(1): 5–14.

Westley, Kieran, Trevor Bell, M. A. P. Renouf, and Lev Tarasov. 2011. Impact Assessment of Current and Future Sea-Level Change on Coastal Archaeological Resources—Illustrated Examples from Northern Newfoundland. *The Journal of Island and Coastal Archaeology* 6(3): 351–74.

Williams, Robert A., Jr. 2012. *Savage Anxieties: The Invention of Western Civilization*. New York: Palgrave Macmillan.

Wilson, Steven G., and Thomas R. Fischetti. 2010. Coastline Population Trends in the United States: 1960 to 2008. Current Population Reports. Washington, DC: US Census Bureau, US Department of Commerce.

Wohlforth, Charles. 2004. *The Whale and the Supercomputer: On the Northern Front of Climate Change*. New York: North Point Press.

Yellowhorn, Eldon C. 1997. Archaeology and the Sechelt Indian Self-Government Act. In *At a Crossroads: Archaeology and First Peoples in Canada*, edited by G. P. Nicholas and T. D. Andrews, pp. 252–65. Burnaby, BC: SFU Archaeology Press.

Yellowhorn, Eldon C. 2002. Awakening Internalist Archaeology in the Aboriginal World. PhD dissertation, Department of Anthropology, McGill University, Montréal, Québec.

Yellowhorn, Eldon C. 2012. Brave New Digs: Archaeology and Aboriginal People in British Columbia, Canada. *The Canadian Journal of Native Studies* 32(1): 87–99.

Zorzin, Nicolas. 2010. Archéologie au Quebec: Portrait d'une Profession. *Archéologiques* 23: 1–15.

Zorzin, Nicolas. 2011. Contextualising Contract Archaeology in Quebec: Political Economy and Economic Dependencies. *Archaeological Review from Cambridge* 26(1): 119–35.

Zorzin, Nicolas. 2014. Heritage Management and Aboriginal People: Relations in a Global Economy—The Contemporary Challenges of Victoria. *Archaeologies: Journal of the World Archaeological Congress* 10(2): 132–67.

Zorzin, Nicolas. 2015. Archaeology and Capitalism: Successful Relationship or Economic and Ethical Alienation? In *Ethics and Archaeological Praxis*, edited by C. Gnecco and D. Lippert, pp. 115–40. New York: Springer.

Index

Albrecht, Glenn *see* dislocation
Alexander, Bruce *see* dislocation
amenity migration 14, 30–5, 37, 84–7, 88, 90, 106; *see also* coastal sprawl; core–periphery; external; population; Sunshine Coast
Anthropocene *see* global ecological crisis; great acceleration; *see also* modernity

Bodley, John 7, 15, 89, 103, 107, 118, 120; *see also* power-elite
bureaucracy 14, 91; *see also* imperialism; instrumentalism; management; resourcism

capitalism 1–7, 18, 50–3, 89–93, 96–7, 99, 102–4, 107–14, 117–18, 120; *see also* disaster capitalism; neoliberalism; neoliberal statecraft
climate change 8, 13–14, 16–17, 19–21, 31, 44, 46–8, 54–5, 58, 63, 84, 99–100, 106, 118–20; impacts on heritage landscapes 23–9; *see also* sea level rise
Club of Rome 98, 111, 115, 121; *see also* critical continuous problems; problematique
coastal change 3, 7, 12–14, 17, 19–37, 54–87, 100, 120; *see also* climate change; coastal sprawl; population; sea level rise
coastal sprawl 14, 17, 19, 27, 30–7, 43; in relation to CRM 47–53; in relation to *shíshálh* First Nation 54–6, 59–63, 76–87; *see also* population
colonialism 49, 108, 117; *see also* capitalism; core–periphery; ideology; imperialism; power-elite
core–periphery 34, 37, 54, 66, 87–9, 94–6, 108; *see also* capitalism; neoliberalism; power-elite
critical continuous problems 98, 121–2; *see also* Club of Rome

disaster capitalism 108, 113; *see also* capitalism; neoliberalism
dislocation 96–8, 120; *see also* slow violence
displacement *see* dislocation

external: external control 42, 85–6, 108; external development 42, 61, 63, 85–6, 94, 107–8; externalities 97, 102

global ecological crisis 1–2, 6–8, 16, 110–11, 117; *see also* great acceleration
great acceleration 6–7, 41, 106; *see also* global ecological crisis
growth, development, and progress 92–3, 99–109, 120; *see also* capitalism; global ecological crisis; great acceleration; modernity

heritage: types of 9–12; critical heritage studies 5; heritage erasure 108, 118; heritage racism 102, 117–18; *see also* Indigenous heritage; management; King, Thomas; site destruction; Smith, Laurajane

ideology 1, 6, 14, 17–18, 88, 90, 93, 99, 101–2, 107–9, 120
imperialism 15–16, 81, 91–2, 101, 108, 110; *see also* colonialism; power-elite
Indigenous heritage: crisis 15–16; Indigenous heritage landscape paradox 111–12; landscapes 10–12; state control of 94–101; *see also* heritage; site destruction
instrumentalism 4, 115; *see also* bureaucracy; management

issue-attention cycle 115–16

King, Thomas 15, 52, 91, 93, 99–101, 109, 113

management 90–2; resource management 92–6; *see also* bureaucracy; resourcism
maritime heritage crisis 1–5, 105–20
modernity 3–7, 91, 95, 97, 99, 103, 106–8, 112, 118, 120, 121–2; four responses to 113–15; *see also* capitalism; growth, development, and progress

Nixon, Rob *see* slow violence
neoliberalism 93–4, 96–7, 107–8; *see also* capitalism; neoliberal statecraft
neoliberal statecraft 88, 93–4, 97, 105, 108; *see also* capitalism; neoliberalism

population 1, 3, 7–8, 13–14, 30–2; in British Columbia 33–4; in Salish Sea 34–5; in Sunshine Coast 60; in Washington State 32–3; *see also* coastal change
power-elite 18, 88–90; *see also* Bodley, John
problematique 98–104; *see also* Club of Rome
psychosocial dislocation *see* dislocation

racism *see* heritage racism
resourcism 92–4; *see also* management
Rogers, Raymond 4–5, 112–13, 120

Salish Sea 1–2, 17, 23, 30, 32–5, 37, 64, 99, 120
sea level rise 1, 3, 12–14, 19–22; impacts on heritage landscapes 23–9; *see also* climate change; site destruction
Sechelt 1, 54, 57, 64–71, 76–87; *see also* *shíshálh* First Nation; Sunshine Coast
shíshálh First Nation 1–2, 15, 17, 54–87; land use plan 56–63
SIT and wait approach 4, 115; *see also* instrumentalism
site destruction 1, 12–13, 17–18, 24–9, 36–8, 49, 61, 65, 75, 78–9, 82–5, 87–90, 93–5, 97–9, 102–20; *see also* dislocation; Indigenous heritage
slow violence 18, 88, 97, 108; *see also* dislocation
Smith, Laurajane 46, 50, 93–6, 101–2, 107
solastalgia 18, 96–7, 118, 120; *see also* dislocation
sprawl *see* coastal sprawl; *see also* population
statecraft *see* neoliberal statecraft
Sunshine Coast 1, 31, 38, 54, 60, 63–5, 84–7; *see also shíshálh* First Nation